PLANTING FRUIT TREES

ALSO BY ROY GENDERS

Bulbs: A Complete Handbook
The Rose
Colour All the Year Round
Covering a Wall
The Greenhouse: For Pleasure and Profit
New Outdoor Tomatoes
Perfume in the Garden
Vegetables for the Epicure

PLANTING FRUIT TREES

The Amateur's Handbook on the Planting and Care of Fruit Trees

by

ROY GENDERS

ROBERT HALE & COMPANY
63 Old Brompton Road London S.W.7

© *Roy Genders 1956*
First published in Great Britain 1956
Reprinted 1973

ISBN 0 7091 4143 2

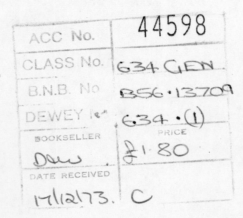

PRINTED IN GREAT BRITAIN BY
REDWOOD PRESS LIMITED
TROWBRIDGE, WILTSHIRE

CONTENTS

	Author's Note	8
	Introduction	9
I	Climatic Conditons—Frost and Rainfall	13

    Value of a variety under varying climatic conditions—The consideration of frost—The protection of fruit trees—The Pear.

II	The Culture of Apples under various Soil Conditions	20

    Apples for a chalk soil—Apples for a cold clay soil—Apples for a wet soil—Apples for a sandy soil.

III	Soil Requirements	25

    To correct an acid soil—Climate and soil fertility—Humus and organic manures—Growing in grass—Providing a balanced diet—Soil requirements of the pear.

IV	Rootstocks and Their Importance	31

    The characteristics of very dwarf rootstocks—Semi-dwarf rootstocks—More vigorous rootstocks—Most robust rootstocks—Pear rootstocks.

V	Pollination and Fertility	37

    Pollination of Cox's Orange Pippin—Biennial bearing—Triploid varieties—Building up a collection of fruit trees—Flowering times—The Pear and pollination.

VI	Planting Fruit Trees	44

    Purchasing reliable trees—Soil conditions for planting—Planting the trees—Staking and tying.

VII	Training Young Trees—Apples and Pears	50

    Training bush and standard trees—Dwarf pyramids—Cordons—Espalier or horizontal form.

VIII	Renovating and Pruning Established Trees	56

    The use and care of tools—Making a correct cut—The functions of Pruning—Renovating old trees—Pruning young trees—Restricting the vigorous tree—Root pruning—Bark pruning—Pears—Grafting.

IX	Care of Young Trees	68

    Building the framework—The established spur system—The Regulated system—The renewal system—Summer Pruning—Biennial cropping—Branch bending—Notching and nicking.

CONTENTS—*continued*

X	HARVESTING AND STORING FRUIT	75

The interest of a fruit store room—Apples—When to harvest—Factors denoting maturity—Storing conditions—Pre-harvest drop—Pears—Harvesting and storing—Plums—Cherries.

XI	THE VALUABLE DUAL-PURPOSE APPLES	81
XII	NEW DESSERT APPLES OF MERIT	85
XIII	APPLES WITH DISTINCT FLAVOUR	90
XIV	DESSERT APPLES FOR ORCHARD AND LARGE GARDEN PLANTING	96
XV	THE RUSSETS	101
XVI	APPLES FOR THE KITCHEN	104
XVII	PEARS THROUGHOUT THE YEAR	108
XVIII	FRUIT GROWING IN POTS AND TUBS	114

Method of growing—Moisture requirements—Suitable varieties—Planting—Culture.

XIX	FRUIT TREES THAT ARE ORNAMENTAL	118

Apple blossom of rich colouring—The richness of pear foliage—Cherry blossom.

XX	THEIR TRAINING AND PRUNING	122

Plums—Spring Pruning—Removal of suckers—Treatment of the fan trained tree—Forming the fan shaped tree—Sweet Cherries—Morello cherries.

XXI	PLUMS AND GAGES: i. *Their Culture*	126

The value of plums in the garden—Growing against a wall—Pollination—Soil conditions—Propagation and rootstocks.

XXII	PLUMS AND GAGES: ii. *Varieties*	132
XXIII	DAMSONS AND BULLACES	139
XXIV	THE CHERRY	142

Their use in the small garden—Soil requirements—Rootstocks—Pollination—Varieties—Acid or culinary cherries.

XXV	SPRAYING FOR PESTS AND DISEASES	148

The Apple; Diseases and pests—The Pear; Diseases and pests—The Plum and Damson; Diseases and pests—The Cherry; Diseases and pests.

	INDEX	155

ILLUSTRATIONS

Facing page

1. Most famous of all Apple Trees. The present owner and her dog beneath the original tree of Bramley's Seedling at Southwell, Notts. — 32
2. Edward VII — 33
3. Crawley Beauty — 33
4. Celia — 48
5. Elton Beauty — 48
6. Opalescent, an excellent dual-purpose apple — 49
7. Exhibition plum, the new Thames Cross — 49
8. Conference pear against a warm wall — 80
9. New pear, Roosevelt. One of the largest and best keeping of all pears — 80
10. Pear, Laxton's Foremost. Ideal for the show bench — 80
11. Upton Pyne, a handsome dual-purpose apple — 81
12. Newton Wonder, a long keeping apple of quality — 81
13. Barnack Beauty, the best dessert apple for a chalky soil — 81
14. Blenheim Orange, a valuable dessert apple for a large garden — 81
15. Three-years-old Dwarf Pyramid apples — 96
16. Cordon apples and 10-tier horizontal pears forming a delightful 'fruit walk' — 96
17. Laxton's Superb, a fine pear for tub or pot culture — 97
18. Two-year cordon apples, Laxton's Epicure — 97

TEXTUAL DIAGRAMS

page

1. Planting a bush apple tree — 48
2. Staking and tying — 49
3. Open centre form — 52
4. Dwarf pyramid form — 52
5. Double cordon — 53
6. Training to the Horizontal form — 54
7. A correct cut — 57
8. A branch carefully removed — 61
9. Building strong fruiting buds — 62
10. Tip bearing tree — 65
11. Spur bearing tree — 65
12. Forming a fruit spur — 69
13. Treatment of an established spur to encourage large fruit — 70
14. Notching and nicking — 74
15. Forming the fan tree — 124
16. A bud for propagation — 130

AUTHOR'S NOTE

IN WRITING this book from the copious notes I have made during twenty-five years of fruit growing, I have kept the amateur gardener constantly in mind. This is his Handbook on fruit tree culture, and is written in the hope that not only will gardeners be more successful with their fruit growing, but that many of the most delicious fruits, now so neglected by the commercial grower because the public buys only a brightly-coloured fruit, will not pass entirely out of cultivation.

In compiling this book, I am greatly indebted to Mr. W. P. Seabrook of the well-known firm of fruit tree growers of Boreham, in Essex, for his kindness in allowing me to make use of his own publications, and for the use of several photographs. Also my thanks to Mr. J. E. Laxton of Laxton Bros. Ltd., of Bedford, to Mr. Alfred Merryweather of Southwell, Notts., and to Messrs. Thomas Rivers Ltd., of Sawbridgeworth, all specialists in the production of fruit trees, for their kind permission to use a number of interesting photographs.

I am also grateful to Doris Gatling for the preparation of the MS., it being no easy matter to arrange the notes of a fruit grower so that they are, as we hope, readable.

ROY GENDERS.

INTRODUCTION

During recent years there have been many books for the commercial fruit grower, but a complete explanation of the arts and mysteries of fruit tree growing has been denied the amateur, possibly in the belief that fruit trees are no longer being planted in private gardens today.

A letter recently received from a fruit enthusiast in Caterham, one of dozens that arrive every week, suggested the writing of this book. It says: "I was on the point of cutting down my fruit trees and using the ground for something else, when I read an article of yours from which it would seem that the cause of the continued failure of the trees is my too-chalky soil. Now I am going to replant with those varieties which you say crop well in a chalky soil, for a garden without fruit trees is quite unthinkable." This confirmed my belief that to the Englishman fruit growing is an integral part of his heritage.

Another writes: "My garden is only large enough to grow two or three fruit trees, so I planted two Cox's Orange Pippin and a Bramley's Seedling, being the recognised best in their respective classes, but although they have been given every attention, they bear little or no fruit." Here complete disregard for the mystery of pollination was the cause of the disappointment and one cannot blame the supplier. When ordering fruit trees without making mention as to why they are being ordered, the nurseryman, always willing to give of his experience, can do nothing but forward the trees without comment.

It has been said that apples and pears were introduced to Britain by the Romans, and certainly it would appear that this was true of the pear. The apple however, seems to have been known to European man since the beginning of time, for in numerous districts fossilized fruits have been found. But to the Romans we possibly owe the first details of the culture of fruit trees, and it would appear that far greater attention was given to the culture of fruit in medieval times than is the case today when the greater part of our population depends almost entirely upon imported fruit with its highly-coloured shining skin, so often devoid of flavour, and which most likely has known long periods in cold storage.

Throughout the centuries gardeners have made their notes in the hope that they may contribute to more successful fruit growing, and as long ago as the beginning of the Wars of the Roses, an aid to better fruit growing appeared, the original manuscript now being in the library of Trinity College, Cambridge. It contains the most excellent advice throughout, but in those days the Englishman had to rely on home grown fruit, and therefore gave much more attention to its culture.

Until the beginning of the 16th Century the Pearmain and Costard were the two apples known, (hence our word costermonger) then an Irishman, Richard Harris, Fruiterer to Henry VIII, brought grafts of ' pippins ' from the continent, planting them on land belonging to the King at Newington in Kent, even then famed for its fruit growing and now the centre of the cherry industry. This was to revolutionise apple growing in England, stock being later obtainable from Richard Harris and used throughout the country, though it is not certain whether ' pippins ' meant a single variety, or several which were new to this country.

It is known that Harris planted russets, now so little grown because we have become slaves of colour rather than of flavour, and because no matter what the soil of our garden, only a brightly coloured fruit must be planted.

" The Pippin burnished o'er with gold," the Golden Pippin, really a russet and grown by Harris may still be found, a tree of which grew in my old orchard in Somerset, the fruit hanging on its leafless branches until the year end, when it may be removed and if carefully stored, will keep until the following summer. It was perhaps of this fruit that Shakespeare's Mr. Justice Swallow said : " You shall see my orchard where, in an arbour, we will eat a last year's Pippin of my own graffing."

What a tragedy it is that we have become so commercialised that almost all the delicious apples known during the 17th and 18th centuries are now stocked by only one or two specialist growers, for the demand is almost negligible. Because nothing but Cox's Orange Pippin, Worcester Pearmain, and Blenheim Orange of home grown apples appear in the shops, the modern gardener believes that they should be grown in every garden to the exclusion of all others. Likewise the Comice pear, so temperamental, and we tend to forget that nearly all our supplies reach us from Italy, where the fruit receives the copious amounts of sunshine it requires. There are other equally delicious fruit far better suited to our climate, but receiving so little publicity that they have become almost extinct.

An example of this is that delicious apple the Devonshire

Quarrenden, known to the Stuarts and cultivated by John Evelyn more than three centuries ago and still widely grown in the West Country. In my opinion it is still the best flavoured apple to follow Beauty of Bath for late August, though perhaps it grows too large for the modern garden. Robert Hogg in the *Fruit Grower's Manual*, published almost a century ago, described it as " one of the best early apples and an excellent bearer "; and Lindley (1830) has said that " cultivated as a dwarf and laden with fruit, it is more ornamental than most fruit trees," with the possible exception of Lady Sudeley.

Another deliciously flavoured apple is the equally old Cornish Gilliflower, mentioned by Parkinson in his *Paradisus*, published in 1629, and of which Lindley wrote " it is the best apple that is known," but away from the West Country it does not crop heavily which may account for its lack of popularity.

I well remember thirty years ago visiting an aged aunt, then living in London in a tiny mews cottage with an equally tiny garden surrounded by a six foot wall. She was a lady born in Herefordshire, where she had spent most of her life, and for her last days she was not to be deprived of her fresh fruit. Along one side of a white-washed wall facing the sun were planted four espalier pears. I cannot remember the names but they were always heavy with fruit during late summer; whilst along a west wall, planted in tubs grew a selection of bush apples, the chief of which I remember was Lady Sudeley ripe by August Bank Holiday. Ideal for tub culture, free of scab, and never dropping its fruit like Beauty of Bath, the skin matures to a rich rosy-apricot colour, the flesh being juicy and of rich flavour. Robert Thompson, keeper of the Horticultural Society Gardens at Chiswick a century ago, described it as being " crisp, rich and sweet, an excellent summer apple " and yet today it is quite neglected. Where space is so limited, could it not once again be planted in tubs, so that one would know what a really delicious English apple is like, eaten warmed by the rays of the August sun.

In my childhood home, the most important room was that fitted up with shelves for storing fruit, apples and pears to last until Lady Sudeley was ripe again in July, and plums too, for several varieties will keep if carefully gathered. The delicious aroma of the room and the fact that we were allowed to help ourselves to the fruit has never been forgotten, and now my own children find equal pleasure in visiting our store-room, a sturdy thatched shed, whenever they wish.

But this pleasure cannot be enjoyed without giving some thought to the art and mystery of fruit growing. This is not a plea to put back the clock, but is a book written for the amateur who would like to know for instance the most suitable fruits to plant on chalk land, the most accommodating, with their compact habit, to plant in a small garden, those to plant in a cold, heavy soil, etc. The mystery and importance of pollination are rarely explained to the amateur though the professional grower knows all about it.

As one travels about England the conclusion is that there is no more typical English scene that fruit trees in blossom in springtime, and laden with fruit in autumn, the cherry orchards of Worcestershire and Kent which conjure up eager anticipation of a new cricket season, of Woolley and Freeman, of Walters and Perks; the cider orchards of Devon and Herefordshire; the russets of East Anglia; pears of Gloucestershire and Surrey; plums of Cambridgeshire and Warwickshire. The townsman who must confine his selection to just two or three varieties in his garden of such limited size, must plant those suited to his soil and climate, and should ignore those varities he sees in his local greengrocer's shop. He requires his trees to bear a heavy crop, and to come into bearing as quickly as possible; he may want to know how to renovate old trees and how to make the best use of a wall for fruit culture. It is necessary to obtain a sound knowledge of all these points and many more before planting fruit trees and in order to avoid disappointment. Then perhaps once again, the quality of home grown fruit may be appreciated as it used to be. Let us start with the selection of the most suitable trees, their rootstocks and adaptability to the soil of one's garden and climatic conditions, and continue right through the culture of each fruit in the hope that in future failure to bear a heavy crop will be greatly reduced.

PART I

APPLES AND PEARS

CHAPTER I

CLIMATIC CONDITIONS—FROST AND RAINFALL

Value of a variety under varying climatic conditions—The consideration of frost—The protection of fruit trees—The Pear.

i. THE APPLE

WHERE APPLES are being planted the selection seems to be governed more by one's favourite variety and its appearance in the shops or on the show bench than for its suitability to the particular soil and climatic conditions of one's garden. Yet there are varieties which have their definite preferences and only by their serious consideration will it be possible to plant fruit trees in the expectation that a satisfactory crop will result. It is little use for instance planting Cox's Orange Pippin above a line drawn from Chester to Lincoln in the hope that it will bear a heavy crop. The Cox's enjoys a warm soil, a porous loam, and a warm climate, and for this reason is at its best in the south.

Requiring quite opposite conditions is that excellent dessert apple, James Grieve, which being a native of Scotland, is at its best in the north, for where the soil is too rich and the climate warm and moist, it suffers seriously from canker. James Grieve delights in cool conditions. Again, Ellison's Orange, does better in a dry district than on the moist west side of Britain.

Much may be learned from the source of origin of the varieties, Bramley's Seedling, first found in a cottage garden at Southwell in Nottinghamshire, crops to perfection in the rich marl soil of the East Midlands. Again, that delicious russet with the crisp, nutty flavour, D'Arcy Spice, nowhere crops more abundantly than in its native Essex, where in 1790, it was found growing in the gardens of The Hall, Tolleshunt D'Arcy. This apple was described by Dr. Hogg in the *Hereford Pomona* of 1880, as being " a dessert apple of first rate character, keeping until May." This is a superb variety for the dry, eastern side of England and its ugly irregular appearance should not be held against its planting in that area.

The variation of the cropping qualities of apples is clearly shown in the report of the South Eastern Agriculture College of Wye, in Kent, for the three years 1948, 1949 and 1950. Of ten varieties selected on fifty farms, James Grieve perfectly happy in the dry climate of South-Eastern England, topped the list with an average yearly weight per acre of 428 bushels (a bushel = 40lb.). Worcester Pearmain occupied second place with 392 bushels, whilst Cox's Orange Pippin was 9th with an average of only 257lbs. If a similar census was to be taken in say, Gloucestershire, the positions of Cox's Orange and James Grieve would undoubtedly be reversed.

THE CONSIDERATION OF FROST

Quite apart from general climatic conditions, the question of situation plays an important part in the cropping of fruit trees. Taken generally, apples are certainly the most hardy of all trees, or top fruits, as they are so often called, but even with apples it is most important to give careful consideration to local conditions. This will be of most importance to those who garden in the North and Midlands where frosts may persist until early June. In the South and West generally, severe frost is rarely experienced after the beginning of May, when the trees will be coming into bloom. The same cannot be said of the South-West where a late frost may be experienced well into May, and which would greatly limit the cropping powers of such frost-susceptible varieties as Cox's Orange Pippin and Bramley's Seedling, the Rolls-Royces of apples in their two respective sections, but both highly difficult to crop consistently well. As soon as their buds begin to burst, they become liable to frost damage and so here again, Cox's especially will prove more reliable if planted in districts where late frosts are not troublesome.

What may be said of frost in general, also appertains to individual gardens, those situated in a frost hollow, or near to a river or stream which tend to be frost pockets. Varieties which may give a good account of themselves in one garden, may prove completely disappointing when planted in a garden susceptible to frost, though perhaps situated only several hundred yards apart from each other. Therefore if frosts are experienced late in the year, no other consideration should be given to the choice of varieties other than deciding whether they will be reasonably resistant to frost. This really means that they will be so late flowering that even the latest frosts do not damage the bloom. Cox's Orange and indeed most dessert apples are out of the question in the frosty garden, though James Grieve, and Worcester Pearmain, the best all-round apples ever introduced, should prove immune, unless the frost is particularly severe. This is why these are two of the most regular cropping varieties for a North country garden, and though James Grieve is one of the earliest apples to come into bloom, it is rarely damaged by frost. These two varieties should not however be planted in definite frost pockets or hollows where frost may be severe.

Where frosts persist late, but are not too severe, the best early maturing apple is Lady Sudeley, raised in Sussex and introduced in 1885. It is possibly the most beautifully coloured of all apples, being of a bright golden-yellow, flushed and striped with scarlet. It is not more widely planted commercially, because it should be eaten straight from the tree when it is soft and juicy but this, however, does not mean that it should not be planted by amateurs.

In bloom about the same time and later than most dessert apples is Laxton's Superb, probably the best of all mid-season apples, which should be planted in Northern gardens instead of Cox's Orange Pippin. The flesh is soft, pure white and exceptionally sweet, and it remains at peak eating longer than any other apple, from mid-November until early February.

For a late dessert apple, in bloom rather later than Superb and of similar appearance is Winter King, introduced by Pope's of Wokenham and renamed Winston, a new apple of great merit. Commercial growers consider it so able to withstand all conditions that if its fruits were slightly larger, it would perhaps be the most widely planted of all dessert apples. The fruit is strongly aromatic and possesses the slightly bitter nuttiness of the russets, but it does require thinning, and to the commercial grower this is not often an economical proposition. With its neat upright habit, Winston is the ideal late maturing apple for a small garden and does well in all soils and with just one or two trees, thinning may easily be done.

For cooking apples, three are of great value where frosts prove troublesome—Newton Wonder, which with its handsome fruit would be the most widely planted of all culinary apples if it was not a biennial cropper; Annie Elizabeth, raised by Harrisons of Leicester, almost as good as Newton Wonder and a completely self-fertile variety; and Lane's Prince Albert, which makes a drooping little tree and remains in bloom over a longer period than any apple tree though is not quite so late flowering as the other two.

Where frosts persist late and severe, in a low lying hollow well away from the coast, one may still grow apples, four varieties blooming very late indeed so that they will miss all frosts, however late in the season. One, Court Pendu Plat, one of the two oldest apples still to be found in England, makes such a small tree of weak habit as to be little planted now, though H. V. Taylor, C.B.E., V.M.H., in his *Apples of England,* describes it as being, " still one of the best of the really late dessert varieties."

Of similar habit is Royal Jubilee, which is also a weak grower though heavy cropper, but the remaining two, Crawley Beauty and Edward VII would be a suitable choice for a cold, frosty garden. These are two excellent apples, heavy cropping and at their best from February to Easter, indeed Edward VII will, if correctly gathered and stored, keep until the earliest maturing varieties are ready again in July. Both are suitable for dessert and for cooking. Edward VII has the excellent Golden Noble for a parent, and was introduced in 1903 by Rowe's of Worcester; whilst Crawley Beauty, was introduced a few years later by Cheal's of Crawley, Sussex. Both crop well in all soils and in all districts, two fine varieties which should be in every garden even where in no way troubled by frosts.

It should be said that where planting these very late varieties, pollination, on account of pollen being so scarce at that time, may in some seasons be insufficient to produce a heavy crop especially if only two or three varieties are planted. It is therefore advisable to plant with them one or two of those varieties which bloom over a prolonged period and which are also resistant to frost damage, such as Lane's Prince Albert, Worcester Pearmain and Laxton's Royalty.

Apples unsuited to areas with a high rainfall—
    Worcester Pearmain, Ellison's Orange, James Grieve, Lord Lambourne, Blenheim Orange, Lord Suffield.

Apples liking a dry, cool climate—
    James Grieve, D'Arcy Spice, Worcester Pearmain, Miller's Seedling, Newton Wonder, Lord Derby, Bramley's Seedling, Sunset.

Apples for a district of high rainfall—
    Grenadier, Allington Pippin, Laxton's Superb, Woolbrook Pippin, Monarch, Cornish Aromatic, Devonshire Quarrenden, Taunton Cross.

Apples for gardens susceptible to frost—
   Crawley Beauty, Edward VII, Royal Jubilee, Court Pendu Plat, Lane's Prince Albert, Forge, Annie Elizabeth, Newton Wonder, Sunset, Laxton's Royalty, Monarch, Mother, Worcester Pearmain.

Apples for a warm district of average rainfall—
   Cox's Orange Pippin, Sturmer Pippin, Gravenstein, Melba, Laxton's Fortune.

PROTECTION OF FRUIT TREES

In this book, which is chiefly intended for the amateur grower with only an average sized garden, it is not possible to make suggestions as to the most suitable land on which to grow apples, rather must "Mohammed come to the mountain" and a selection be made to suit one's existing garden. No professional grower would plant in a frost hollow, therefore mention is rarely made of suitable varieties for such land. But though the gardener must do the best with the situation and climate at his disposal, it must be said that the matter of pollination is all-important and the maximum of pollination will only take place where the insects are given some protection from cold winds. This is chiefly why fruit trees crop so abundantly in a walled garden where the atmosphere is calm and warm, and much more helpful to bees and other insects than where the trees are continually swept by prevailing cold winds.

So if at all possible select a position where the trees receive some protection from strong winds, though it is even more necessary for them to receive their fair share of summer sunshine, so essential for ripening the fruit. Plant pears in a position of full sun for this is more important to growing a top quality pear than an apple, but give the apples as open and light a position as possible. Planting in the almost complete shade of a building, or too close to matured trees which have been planted for wind protection, or for their ornamental value, will cause the fruit trees to become drawn and weakly and they will never bear an abundant crop.

If planting a small orchard in a wind swept district it will be advisable to use the Myrobalan or Cherry Plum, or the Damson, for a windbreak. Both are strong and their fruit is valuable for preserving. Neither do they deprive the soil of moisture and nourishment as will the poplar, thuya, or most other hedging plants.

ii. THE PEAR

From the shores of the Black and Caspian Seas and from the Eastern Mediterranean Seaboard, the pear is to be found growing wild and crops in abundance, and it is from these areas that it reached Italy and Greece several centuries B.C., and later Britain, with the Roman Invasion. Whereas the apple is European, the pear is of Asiatic origin, coming into bloom before the apple and requiring, as would be expected, a warmer climate. The area of good pear crops in Britain is confined to the area south of a line drawn across England from Worcester to The Wash, only the very hardiest of pears being grown north of that line. Even in the South the pear growing regions are of East Anglia and Kent, and in those counties through which flows the River Severn. The chalk soils, often dry and shallow, of the mid-Southern counties of Wiltshire, Hampshire, Oxfordshire and West Sussex, though suitable for apples, rarely grow good pears which must have a rich moist loam.

The pear must have its roots in a warm moist soil, and its head in dry sunshine. The West Country in general often produces scab, the North Midlands and North Country is too cold, whilst the Southern chalk brings about such a deficiency from iron that the trees make but limited growth and bear only small crops. Away from Kent, East Anglia and the Severn Valley, the pear may be considered the most difficult of all fruits to grow well. An English grown apple will stand comparison with any in the world, only rarely does the English grown pear.

But the pear may be cropped more satisfactorily than is so often the case if the same rules as to climate are followed as for the apple. In the colder North and North Midlands, there is no reason why the most hardy varieties should not be planted. It is quite useless to plant Doyenne du Comice and Roosevelt, as delicious as they may be, with the expectation that they will crop well, But the old Jargonelle, Beurré Hardy, Catillac and Durondeau, will all bear quite well at almost 1,000ft. above sea level and in a sheltered position, Laxton's Superb, Dr. Jules Guyot and William's Bon Cretien, may be added provided they are given the protection of a wall.

Each of these pears are not only hardy, but by coincidence all flower late, and so miss reasonably late frosts even in the colder districts. Given a sheltered garden, and preferably a wall for protection, they may be grown as far north as Lancashire and Derbyshire and even further north in suitable gardens.

The same consideration should be given to the planting of indi-

CLIMATIC CONDITIONS—FROST AND RAINFALL

vidual varieties as with the apple, for although the pear enjoys a warm soil, a too moist climate will cause outbreaks of scab to which the trees will eventually succumb. Both Clapp's Favourite and Pitmaston Duchess, heavy bearers and which make large, vigorous trees, suffer badly from scab in the moist climate of the West.

Whereas there are numerous apples which will yield heavily in a chalky soil, there are no pears which will tolerate such a soil, likewise in the soot-laden atmosphere of our cities the apple is able to bear heavily whilst the pear will bear only spasmodic crops.

Frost pockets must be avoided for the pear blooms earlier than the apple and even the latest of all to bloom, Catillac and Winter Nelis will not be late enough to miss late frosts as will apples Crawley Beauty and Edward VII. The choice then with pears is much more limited than with apples, but as the fruit is so delicious, everything possible should be done to plant those varieties which will ensure some degree of success under prevailing conditions.

CHAPTER II

THE CULTURE OF APPLES
UNDER VARIOUS SOIL CONDITIONS

Apples for a chalk soil—Apples for a cold clay soil—
Apples for a wet soil—Apples for a sandy soil.

ALL APPLES are happiest planted in a soil that is well drained and of a rich heavy loam, but not all gardens are lucky in this respect, quite often the trees having to be planted in a cold, clay soil especially where taking over new property. The commercial grower would select more suitable ground before embarking on his planting, selecting the land to suit the fruit. The amateur however, must make the best use of what is available, it not being possible to terminate one's work and move into a more favourable district just because the household demands its home-grown fruit. But whilst most pears must be given a warm soil, and cherries one of a chalky nature, there is a wide selection of apples suitable for most, and including the ' difficult ' soils, with which the amateur gardener frequently has to contend. These will be:—

(a) Soils of a chalky nature, which are generally shallow and likely to lack moisture during a dry period;
(b) Those of a cold, heavy nature;
(c) Those wet and badly drained; and
(d) Those very light and sandy.

APPLES FOR A CHALKY SOIL

By far the most difficult soils are those (a) of a chalky nature chiefly to be found in parts of Yorkshire, Derbyshire and across the South Midlands from Wiltshire, North Hampshire to London. Also, with the exception of Derbyshire, being in an area of low rainfall, lack of moisture about the roots is often experienced at a time when the fruit is maturing, and needs the necessary moisture to make reasonable size. This is the reason for small, tasteless fruit.

CULTURE OF APPLES UNDER VARIOUS SOIL CONDITIONS

The trouble may be partly overcome by manuring and by deep digging. The use of rape seed for green manuring, sowing the seed in April and digging in when several inches high will help to add fibrous humus to the shallow soil. The additional care with planting the trees will also help to ensure that the trees do not suffer from drought, but most important of all it is necessary to plant those varieties which have proved themselves capable of bearing a heavy crop in a chalky soil, for quite apart from the question of moisture, some varieties show a liking for chalk, whilst others definitely do not. The most important of these is Barnack Beauty, introduced in 1880 by Messrs. Browns of Stamford, Lincs., and which received an Award of Merit in 1899. Yet this is an apple which has never become widely grown, because it is one of the few which is much happier in a dry, chalky soil than in a rich loam. If more amateurs knew of its merits when planted in chalk, many disappointments would be prevented, for here it bears heavily.

The tree is a strong grower and like Worcester Pearmain, it is a tip bearer; that is, it bears its fruit on the tips of the twiggy wood, rather than forming close spurs as does Cox's Orange. For this reason it is a tree for a good sized garden, and it not suitable for dwarf tree culture, e.g. cordons. It is a handsome fruit, greeny-yellow flushed bright crimson on the sunny side, and it will store until April. Indeed it is not at its best until March, when it is juicy and sweet.

Almost as good is Barnack Orange, the result of a cross from Barnack Beauty and Cox's Orange. It is a new apple, at its best at Christmas, and like Barnack Beauty never crops well away from chalk. Another is St. Everard, ready early in September, when it is quite the best of the early apples, but away from chalk it crops so poorly and makes only a small, weak tree.

Making a tall, spreading tree and therefore better for a large garden is Gascoyne's Scarlet, raised in the chalk soils of Sittingbourne, and like Barnack Beauty, it crops well only on chalk. It bears a most handsome fruit, the skin being of palest cream, flushed with vivid scarlet, the sweet flesh also being coloured with scarlet. The fruit is equally good for cooking or for dessert. Another apple that crops well in a shallow, chalky soil is Charles Ross, a delicious and most handsome apple, but it badly deteriorates if kept after November. It cooks well. The tree is of vigorous, spreading habit, and so better in a country than a town garden. Here then are several good apples which will bear well on chalk.

APPLES FOR A COLD CLAY SOIL

Soils of a cold, heavy nature (b) which are so often to be found in towns and in certain areas of the Midlands will be quite unsuitable for Cox's Orange, and it is little use planting any variety which has not proved suitable under cold conditions. The same varieties should also be given preference for planting in cold, exposed Northerly gardens and those situated in the North Midlands, and in parts of Scotland and Northern Ireland. Grow Cox's Orange in Dublin, Allington Pippin in Belfast.

Allington Pippin is a fine dual-purpose apple, widely grown in the Fens, Lincolnshire and East Yorkshire, but it blooms early and should not be planted in a frost pocket. Perhaps a better variety, if a dessert apple is required is Adam's Pearmain, possibly the hardiest of all dessert apples, and one of the very best for a cold, exposed garden. Why is it so little known? Lindley wrote, "Its merit consists in it being a healthy, hardy sort, a particularly free bearer, extremely handsome, a good keeper, . . . acid and sugar being so intimately blended as to form the most perfect flavour." Added to this is its tolerance of cold, clay soils and what more is required of it? The habit is compact, whilst the fruit is at its best after Christmas.

For October, for a cold, heavy soil King of the Pippins, possibly 200 years old, is an interesting apple. The fruit is oblong and of a bright orange colour, the flesh being hard, dry, aromatic, and with a pleasant almond flavour. It is extremely hardy and is a heavy bearer.

For a culinary apple, ready in September, none is hardier than Pott's Seedling, raised in 1849 at Ashton-under-Lyne. It makes a compact tree and bears heavily. For late winter use, Newton Wonder, raised at King's Newton, Derbyshire, is as hardy as any apple and quite happy in a heavy clay soil. For size, shape and colour it is one of the most handsome of all apples and should be in every garden irrespective of soil and climate.

One for the City stockbroker who enjoys a home-grown apple, but who prefers to leave the trees entirely to nature once they are planted, is Herring's Pippin. Extremely hardy in a cold, clay soil, this apple will continue to crop well if never pruned, sprayed nor mulched. It is foolproof, and yet bears a huge crop of handsome green and crimson fruit which possess a strong, spicy flavour, useful either for cooking or dessert. Like Newton Wonder, this too, should be in every garden especially in the North, planted with Edward VII.

Something may be done to bring a clay soil into as satisfactory a planting condition as possible by incorporating drainage materials, such as grit, crushed brick, and coarse sand, but first dig in deeply some caustic or unhydrated lime. This should be done during the early part of winter, the action of the lime as it decomposes also breaking up the clay particles in the soil. Then work in the drainage materials and finally some humus, such as strawy farmyard manure or straw decomposed by an activator, together with a small quantity of peat. Allow the soil to settle down and plant in March.

APPLES FOR A WET SOIL

For a wet or moist soil (c)—which does not mean a cold, clay soil, but rather where land is low lying, possibly of a loamy nature, and generally containing a high percentage of sand—again little can be done except where planting a small orchard, to dig drains or trenches between the trees. Those varieties prone to scab should be avoided, but Lord Derby, Grenadier and Monarch, all excellent apples, are outstanding in their ability to tolerate a wet soil, and even cold conditions. These varieties are described in Chapter xviii, but I should like to mention in particular the performance of Grenadier in my Somerset orchard. In a low lying part to which moisture drains, the trees are frequently covered with water, often 6–8 inches up their stem, being standard trees. Following heavy rains the tree roots are often submerged for several consecutive weeks, and yet year after year, they bear heavy crops of the most immense apples ever seen. It should be said that Grenadier is the best of all pollinators for Bramley's Seedling, but whereas this fine apple will, like Grenadier, tolerate cold, heavy soils, it does not like wet, low-lying ground where it would also be highly susceptible to blossom frost damage.

Of dessert apples suitable for a wet soil, two will be found reliable croppers, Laxton's Superb, which makes only a small tree and comes quickly into heavy bearing, its handsome fruit being at its best in December; and Sam Young, a little known variety of Irish origin and not even mentioned in H. V. Taylor's comprehensive *Apples of England*. Robert Hogg in his *Fruit Manual* describes it as being, " a first-rate little dessert apple, in use from November till March." It is a russet, cropping heavily in a damp soil, and though the fruit is small, it is one of the most richly flavoured of all apples.

APPLES FOR A SANDY SOIL

Lastly, light sandy soils (d) which are generally well drained and which may be made more retentive of moisture by incorporating plenty of humus in the form of decayed stable manure. Light soils generally prove deficient in potash, which is described in the following chapter. The two best dessert apples for a light soil, yet not in an area of too heavy rainfall, are Worcester Pearmain and Ellison's Orange. Worcester Pearmain, introduced by Smith's of Worcester in 1873 is an indispensable apple to the commercial grower, an excellent pollinator for Cox's Orange, being a heavy and reliable bearer, with its fruit a rich crimson colour which sells on sight. Unfortunately for the amateur's garden, it is of vigorous habit and is a tip bearer and so is not suitable for dwarf forms. It is a gross feeder and requires plenty of room, and so is better suited to a large garden or small orchard.

For a small garden, Ellison's Orange, which comes quickly into bearing, its fruit maturing by October, will prove reliable in a sandy soil. It is a heavy and consistent cropper, and if it has a fault it is that its highly aromatic fruit takes on an aniseed flavour if stored too long. It should be consumed before November 1st.

A long keeping apple for a light soil, at one time widely planted throughout Sussex, is Forge, which Hogg describes as "the cottager's apple *par excellence*." It is a huge and consistent cropper, the greasy skinned fruits keeping well into the new Year. It makes a compact tree, is scab free and extremely resistant to frost.

Apples for a chalky soil—
   Gascoyne's Scarlet, Barnack Beauty, Barnack Orange, Charles Ross.

Apples for a cold, clay soil—
   Allington Pippin, Adam's Pearmain, King of the Pippins, Pott's Seedling, Newton Wonder, Herring's Pippin.

Apples for a wet soil—
   Lord Derby, Grenadier, Monarch, Laxton's Superb, Sam Young.

Apples for a light, sandy soil—
   Worcester Pearmain, Ellison's Orange, Forge.

CHAPTER III

SOIL REQUIREMENTS

To correct an acid soil—Climate and soil fertility—Humus and organic manures—Growing in grass—Providing a balanced diet—Soil requirements of the pear.

i. THE APPLE

APPLES LIKE a soil containing both humus and nutrition. It is possible to give one without the other with the result that a correctly balanced soil will not be obtained. We have seen that where planting in a heavy soil, some straw manure is necessary to prevent the soil particles becoming too compressed. This would deprive the roots of the air so necessary to bring about bacterial action of the soil, without which the trees are unable to derive full benefit from manures. Humus is also necessary in both a light and shallow soil, but for a different reason. Here humus is provided to retain the maximum amount of moisture in the soil throughout summer, and though certain varieties will crop reasonably well in a light, poor soil, they will crop much better where both humus and a balanced diet is provided. Newton Wonder and Lord Derby, will bear a heavy crop in light soil, but will only do so if humus and the requisite potash is present.

TO CORRECT AN ACID SOIL

Besides the necessity for providing humus to counteract any deficiency in the soil, it is also important to ensure correct soil fertility. This means a balanced supply of the necessary foods to be made available to the trees over as long a period as possible, preferably for the lifetime of the tree.

As the soil of many town gardens is of an acid nature, due to deposits of soot and sulphur over a period of years, most town gardens will respond to a dressing of lime. This does not mean a heavy dressing unless certain varieties (e.g. Barnack Beauty and Gascoyne's Scarlet) are to be planted, and which will benefit from a heavy application. As the apple requires magnesium in the soil, lime is best given in the form of magnesium limestone, or magnesium carbonate, and this should be applied in the early winter at the rate of 2lbs. per square yard. If planting a small orchard, the soil should be tested before applying the limestone and this will only be necessary if the pH value shows an acid reaction below 6.0. Soil which is of an acid nature cannot fully convert the food content for the nourishment of the plants.

CLIMATE AND SOIL FERTILITY

Fruit trees require food for them to grow, and to make new wood throughout the life of the trees. The trees also require a balanced diet to enable them to bear a heavy and consistent crop, but as to quantities of food required much depends upon soil, situation and variety. Apples growing in a light soil will require considerably more potash than those in a heavy loam, whilst those growing in an area of high rainfall, which will tend to wash the minerals from the soil, will need heavier manuring and more careful attention to the plant's requirement in this respect. Variety too, plays an important part, for the most vigorous growers e.g. Bramley's Seedling and Blenheim Orange, will become too vigorous and make an excess of wood and leaf if given the same amount of nitrogen needed for Sunset or Adam's Pearmain. In general, dessert apples require more nitrogen than the cookers, but as it is potash rather than nitrogen which gives colour to an apple, those noted for their high colour e.g. Worcester Pearmain and Charles Ross, may be kept on a low nitrogen diet. Likewise those varieties which tend to be troubled by scab, for nitrogen tends to make excessive soft growth, and scab will be more prevalent.

Again, trees growing on the Eastern side of Britain, where growth is much slower than on the West, will need additional nitrogen. It is therefore important to consider each tree and local climatic conditions before applying the manures, remembering that cooking apples will be of a more vivid green when given additional nitrogen, whilst dessert apples will show richer colourings of scarlet and orange if given liberal supplies of potash.

HUMUS AND ORGANIC MANURES

When preparing the ground for planting new trees, dig in deeply some farmyard manure, or straw compost, possibly containing decayed vegetable waste, and which has been rotted down with an activator. Possibly poultry or pig manure will have been incorporated. An average dressing is 10 tons to the acre, or roughly

SOIL REQUIREMENTS

half a barrow load per square yard. This manure will contain nitrogen, potash and phosphates, it is slow acting which is advantageous to fruit trees and will also improve the structure of the soil. It is important to dig it deeply in and mix it well into the soil; manure however, should not be packed around the roots of the trees.

Those who live near the coast could use seaweed as an alternative, or even fish waste. For those who live in the Industrial North, shoddy, rich in nitrogen, and wood ash, rich in potash, may be used together and in place of farmyard manure. As apples also require small quantities of phosphorus, for otherwise the fruit will tend to remain small, 2ozs. per square yard of either bone meal, or steamed bone flour should also be worked in. To prevent any waste, for all manures are now expensive, a square yard of ground should be marked out and prepared exactly where the trees are to be planted.

Preparation of the ground should be done whenever the soil is friable during winter, December being an ideal month, for then the soil will be weathered by winter frosts, and will have settled down to enable planting to take place in early March. This is generally the best time for heavy soils, though any time during winter will suit a light soil.

The following are organic manures to provide nitrogen:

Manure	Content	Action
Dried Blood	12%	Rapid, but lasting
Feathers	8%	Very Slow
Hoof and Horn Meal	12%	Slow
Shoddy	7–14%	Slow
Compost	0.3%	Slow
Fish Waste	4–10%	Very Slow
Fish Manure	10%	Rapid
Bone Meal	4–8%	Slow
Seaweed	0.4%	Slow
Spent Hops	4%	Medium

Only slow acting manures should be given when the trees are first planted, but as most apples and certain varieties in particular, require readily available potash, either sulphate or muriate of potash, inorganic manures, containing about 50% pure potash, may be given at planting time at the rate of 2–3 ozs. per square yard. Only where planting trees on Rootstock Malling IX is it not necessary to provide this additional potash.

After planting, a summer mulch of farmyard manure placed around each tree is all the attention necessary as far as soil fertility is concerned, for at least a couple of years.

GROWING IN GRASS

There has always been much controversy as to whether one's fruit trees should be planted into pasture (grass covered ground), or into ground where the grass has been removed and which is possibly used for vegetable culture. Some compromise, by making large circles round the tree stems after planting.

It may be said that young trees will suffer serious nitrogen shortage if planted directly into grass, and older trees will be deprived of much valuable nitrogen, unless the grass is either grazed or kept cut short. Weed infested grass, allowed to grow tall will prove highly detrimental, especially in areas where growth is perhaps slower than normal. Grass which is grazed will use up little of the nitrogen in the soil and this will be replaced by the droppings of the animals. Grass which is cut constantly and left as a mulch will not only remove very little nitrogen, but will return valuable potash to the soil. Where grass is grazed some additional potash will be necessary, quality and colour being the deciding factors.

But always a balanced diet is just as essential for fruit trees as for humans, nitrogen and potash requirements going hand in hand, an excess of one and shortage of the other proving detrimental to plant growth, whilst to a lesser extent phosphates and magnesium are also essential for a balanced diet. It should be said that where being planted in grass which cannot be kept short, the trees will benefit from the removal of a large ring of turf from around the base, and over the exposed soil the mulch is given each year.

In districts of low rainfall, or where the soil is shallow, often overlying chalk trees should never be planted in grass, for every drop of moisture will be needed by the trees. Here a mulch will prove most important and if farmyard manure cannot be obtained, cover the ground with decayed leaves, peat or composted straw, and this should be applied in May, before the moisture evaporates from the soil. Fresh straw or sawdust should not be used for this will utilise a considerable amount of nitrogen as it is undergoing the process of composting, and which is needed by the tree.

PROVIDING A BALANCED DIET

When the trees have been in their new quarters for two or three years careful check should be made on growth. If the trees are growing in an average loamy soil, and in a district of average rainfall, new, or extension growth, will be between 1–2 feet each season. If less than 12 inches of new growth is made, then there will be need to provide a nitrogenous dressing, and the same may be said of established trees. Remember that certain varieties possess naturally a most vigorous habit, and these should be considered when determining nitrogenous requirements. The necessary nitrogen may be given by an application of sulphate of ammonia, raking round the tree about 2 ozs. during late March when growth commences. More established trees may require double this amount but it is much better to give a small quantity each year rather than a large amount one year, and none for several more years. Trees growing in long grass should be given from 3–4 ozs., for here nitrogen deficiency will generally be greater.

A deficiency of potash, phosphorus and magnesium, will not become apparent until the trees come into regular bearing, though lack of potash may be observed by the foliage turning brown and becoming crinkled at the edges. When the trees is bearing, the fruit will be small, of poor colour and will not keep as it should. There will also be a dearth of fruit buds, and so the crop will be small. Trees starved of potash wil gardually become bare of foliage, and will die back altogether. But potash must only be given where nitrogen is present, or with it, for it is essential to maintain a balanced diet. Potash in the soil will also release the necessary phosphates, so important for size of fruit, and thus it is only rarely that phosphates are artificially given by themselves.

Certain varieties show greater tolerance to potash deficiency than others, Worcester Pearmain is one which is very tolerant, whilst Lord Derby, Newton Wonder, Grenadier, Cox's Orange, Miller's Seedling and Beauty of Bath, all require liberal potash supplies especially on a light, sandy soil. It may be said that dessert apples require more potash than cookers, and most will require double the quantity on a light soil as those growing in a heavy loam.

Potash is best given in the form of sulphate of potash, using 1 oz. per tree, raked into the soil in spring. Too heavy applications must not be given for this will cause magnesium deficiency, only recently realised as being due to the use of excess potash. The trouble is frequently encountered where the soil is light and has received yearly potash applications, thought to be necessary. A tree deficient of magnesium will be observed by the foliage turning a pale green colour, and with purple blotches appearing about the centre ribs. Also the new wood will shed its foliage too soon, before the end of summer. If left uncorrected, the trees become stunted and bear small crops.

The trouble may be prevented by an occasional dressing with magnesium limestone, especially where the soil tends to be sour, but a more rapid corrective will be to spray the foliage in mid-summer (June), with a solution of magnesium sulphate (Epsom Salts) made by dissolving 1 lb. to 5 galls. of water, the magnesium being absorbed through the leaves.

An excellent method of maintaining the health of a young plantation is to lightly fork fish manure around each tree in December. This contains a high percentage of nitrogen, potash and phosphates, in addition to being of an organic nature. Farmyard manure is then used as a summer mulch.

ii. THE PEAR

Whilst the manurial requirement of the pear is similar to that of the apple, the greatest use should always be made of farmyard manure. The greatest requirement of this fruit are moisture, supplied by the humus content of stable manure; and nitrogen, pears demanding little else. If these are always available and a warm climate provided, then good crops of pears may be grown. Pears will not respond in anything like the same way as will apples to applications of inorganic fertilisers, though where the humus content of the soil is considerable, light dressings of sulphate of ammonia, to provide additional nitrogen, may be given at the rate of 2 ozs. per tree in April. Fish manure and bone meal are also valuable manures, in fact, all the organics rich in nitrogen, and one of the most renowned pear growers in Cambridgeshire insists that the secret of his success is providing his trees (Conference) with a heavy annual application of either farmyard, or some other organic manure. This is best given during Marsh, and dug into the soil round the trees, when the ground is in a suitable condition. Trees growing in grass should receive this manure by way of a heavy mulch.

Even more important is to ensure that those trees growing against a wall, as do pears more than apples, receive sufficient moisture and nitrogen. Lack of both is one of the greatest causes of lack of wall trees bearing heavy crops. Farmyard manure, shoddy or fish manure should be lightly forked around the trees in early winter, and this should be followed by a heavy mulch in May, before the moisture has commenced to dry out.

Pears, planted in a limestone soil may occasionally show symptoms of lack of iron, far more prevalent amongst pears than with apples. This may be cured by drilling a small hole into the main trunk of the tree, and inserting a tablet containing 1 gramme of ferrous sulphate. An older generation of gardeners hammered a large nail right through the trunk and the effect was similar.

CHAPTER IV

ROOTSTOCKS AND THEIR IMPORTANCE

The characteristics of very dwarf rootstocks—Semi-dwarf rootstocks—More vigorous rootstocks—Most robust rootstocks—Pear rootstocks.

i. THE APPLE

THE QUESTION of rootstocks might seem a complicated business, yet it is one of the utmost importance to the amateur, for the habit of each variety will greatly depend upon the planting of a certain rootstock. These rootstocks, upon which the grafts or buds are made, may be divided into four types.

(a) Those of very dwarf habit;
(b) Those of less dwarf habit;
(c) Those more vigorous; and
(d) Those of very robust habit.

Upon which stock is being used will depend the planting distances and ultimate weight of the crop.

(a) VERY DWARF ROOTSTOCKS

On the very dwarf rootstocks, Malling IX, the trees will come more quickly into heavy bearing than on any other rootstock. It is that often used for cordon and pyramid trees which are required to be of less vigorous habit than bush and standards, but it may also be used for bush trees, which are required for a small garden and which are expected to remain reasonably dwarf and yet come early into bearing. In other words, trees on this rootstock fruit abundantly at the expense of making wood, though they have a tendency to burn themselves out after thirty years fruiting, unless carefully tended throughout their life. This means:

(a) Regular attention to pruning, though this will not present much of a problem with the trees making little new wood;
(b) Care in keeping the ground clean; and

(c) Providing a regular balanced diet.

They also need careful staking during the first years after planting for their root action, as may be expected, is not vigorous, and the trees easily blow over.

Bush trees on this rootstock should be planted from 10–12 ft. apart, pyramids 6–8 ft. and cordons 2 ft. apart, but much depends upon the natural habit of each variety. Beauty of Bath for example makes a spreading tree and should be given wider spacing than say, Adam's Pearmain; whilst the more vigorous varieties such as Bramley's Seedling, Blenheim Orange and Miller's Seedling should never be planted on this dwarf rootstock, for weight of foliage and fruit would most likely prove too much.

Trees on Malling IX, which are planted closer than those of other rootstocks, and coming into heavy bearing sooner than any other type, will ensure the largest weight of fruit from a small garden, in the quickest possible time. But it must be said that these trees will never yield much more than a bushel of fruit (40 lbs.) at whatever age they reach, whereas Bramley's Seedling and Newton Wonder, on Malling II will yield up to 10 bushels when in full bearing, and Blenheim Orange and Worcester Pearmain about half that weight.

A more recent dwarfing rootstock is the Malling-Merton, MM 104, which has so far shown a greater resistance to woolly aphid than Type IX, and has also given heavier crops, especially with Cox's Orange Pippin and Ellison's Orange. At recent trials of the East Malling Research Station in Kent, 11 year old Cox's Orange produced 461 lbs. per acre on this rootstock compared with 382 lbs. on Type MIV, which is similar to MIX.

Another new rootstock MM 106 gave in comparison only 257 lbs. on the same trees, planted in the same heavy loam, but this stock has proved a much heavier bearer when used in light, sandy soil. With its much better anchorage, and being a more satisfactory propagator for the nurseryman, MM 104 looks as if it will supercede all other dwarf rootstocks in the years ahead.

(b) SEMI-DWARF ROOTSTOCKS

Malling IV has for some time been the recognised rootstock with a semi-dwarfing habit, and over a period of the first 25 years, trees will bear larger crops than with any rootstock, but like MIX the trees root badly and require to be well staked. This rootstock suits Cox's Orange better than any other and is generally used entirely for this purpose, but it is safer to plant in a position sheltered from strong winds. From the latest results of the new

Most famous of all apple trees. The present owner and her dog beneath the original tree of Bramley's Seedling at Southwell, Notts.

Edward VII

Crawley Beauty

TWO VALUABLE LATE APPLES FOR A FROSTY GARDEN

ROOTSTOCKS AND THEIR IMPORTANCE

MM 104 it would appear that this rootstock will eventually supercede both MIX and MIV.

Planting distances should be increased by 2 ft. for bush trees where using MIV, as in their first 10–12 years, the trees will make rather more growth than on MIX. Apart from the same need to stake securely they will prove more suitable for orchard culture, not requiring quite such detailed attention as those on MIX.

(c) MORE VIGOROUS ROOTSTOCKS

For large garden or orchard planting MI and II have been most successful during the past twenty years or more. Trees on MI will come more quickly into bearing than those on MII, and it has been observed that MI is better suited for planting in a district of high rainfall and soil moisture, and so has been propagated considerably by West Country nurserymen. It does not suit Cox's Orange and with many varieties in a dry soil it has not resulted in anything like such heavy crops as from trees on MII, which may be said to be the most successful of all rootstocks for commercial planting. It produces a tree of good size which it builds up gradually, at the same time increasing its cropping so that there is a balanced, long lived tree, which over 50 years or more will bear a heavier weight of fruit than from any other rootstock.

As the trees, especially of the most vigorous varieties will require a spacing of 20 ft. where the soil is of a heavy loam, this is not a suitable rootstock for a very small garden, but much depends upon whether one wishes to enjoy a heavy crop during the first 20–25 years of the tree's life with diminishing crops later, or more regular bearing over 50 years or more.

Trees on MII root deeply and rarely require staking, and being able to search widely for their nourishment are good orchard trees, able to withstand drought and do not require such meticulous attention as to feeding and upkeep. They will however, require much more attention as to their pruning. Certain varieties, such as Winston and Sunset, which are of only semi-vigorous habit in comparison with Blenheim Orange or Newton Wonder, would find MM 104 more suitable, for these less vigorous varieties will give a greater yield per acre on a less robust tree, and planted half the distance apart of that allowed for MII rootstocks. The amateur can draw his own conclusion, possibly planting six trees 10 ft. apart on MIX or MM 104, or on the same area of ground, three trees 20 ft, apart of MII or the new rootstock MMIII. But whereas the six dwarf trees may yield around 150 bushels of fruit over the first 30 years of their life, the three trees on the vigorous stock should yield 300 bushels, though the greater weight will be given between 20–30 years of age. For an orchard or large garden, the more vigorous rootstock is to be recommended; for a small garden, the more dwarf rootstocks.

The new MMIII in the East Malling Trials has cropped more heavily than MII, a combination of Cox's Orange, Jonathan and Ellison's Orange yielding an average of 348 lbs, from 11 year old trees, compared with 273 lbs. on MII, and with there being little difference in the size of tree, it would appear that MMIII will become widely used in future years. It is also remarkably free from woolly aphid.

(d) MOST ROBUST ROOTSTOCKS

The two most vigorous rootstocks yet produced are the old MXVI and the new MM 109, the latter however may be said to come somewhere between groups c. and d. for vigour. Both make extremely large trees, as bush or standards, and require planting between 20–30 ft. apart. This stock is generally used for standard orchard trees and will take from 15–20 years before coming into heavy bearing. At from 10–12 years of age it bears only half the crop produced by MIX or MM 104 at the same age, and very much less than given by MMIII or MII. These extremely robust rootstocks are of little use for the amateur's garden, but may be used for those culinary varieties which make rather sparse growth e.g. Lord Derby and Grenadier, and which will respond favourably to the additional vigour.

MM 109 has so far shown a high resistance to drought, more so than either MII or MMIII, and for this reason is being planted, and might well be more so, in East Anglia, also where the soil is shallow and overlying chalk, so often a dry soil. No comment can be made on Barnack Beauty or Gascoyne's Scarlet, used on MM 109, but one would imagine this rootstock to give heavy crops planted in a dry, chalky soil.

A new rootstock of great vigour, MXXV has shown that trees will come into heavier bearing at an early age than those on MXVI or MM 109 in spite of it making a large tree, but it has been troubled by woolly aphid, and it is early yet to acclaim it as being an advance over others of robust habit.

All this may, to the amateur sound most complicated, but it is advisable to consider the part played by each rootstock, not only on the weight of crop, but on the habit of each variety, so that the correct choice may be made for the size of one's garden. To put it more clearly:

(a) Very Dwarf Rootstocks—MIX, MM 104.
(b) Semi-Dwarf Rootstocks—MIV.
(c) More Vigorous Rootstocks—MI, MII, MMIII.
(d) Most Robust Rootstocks—MXVI, MXXV, MM 109.

ii. THE PEAR

Though rootstocks do play a part in the habit and cropping of pears, the choice is not nearly so wide as for apples. This is due partly to the fact that pears respond well to the already existing rootstocks, and to the fact that with this fruit being of less importance in comparison to the apple, nothing like as much interest has been taken in the pursuit of experiments on other rootstocks, which might possibly prove more satisfactory.

As long ago as 1667, Merlet, the French horticulturist recommended working pears on the quince stock, as producing far better fruit than when worked on the then commonly used hawthorn. At the same time Le Gendres made reference to this in detail in his *Manual of the Cultivation of Fruit Trees,* noting that the " graft swells equally fast with the stock, which none of the others do." The quince stock is still used, Quince A and B may be compared with Type I and II in apples, producing a large tree and taking longer to come into heavy bearing than the more dwarfing stocks. These two stocks are usually employed for bush trees, for orchard or large garden planting, and in good pear growing districts should be planted 15–18 feet apart.

Quince C is that most frequently used, this being similar to Type IX for apples. It makes only a dwarf tree and comes quickly into bearing, the weight of fruit, when in full bearing almost bringing tree growth to a halt. This is the stock used for pyramids and for the cordon and espalier form. Pears which are normally slow to come into bearing, such as Doyenne du Comice and Buerré Hardy should always be worked on this stock.

Where a large standard tree is required, pears are worked on the pear stock and this stock is also used for those of weak habit.

With several varieties it has been found that they will not ' take ' to the quince stock and need to be what is termed by the nurseryman, " double-worked." This means that a variety known to ' take ' well on quince, such as Beurre d'Amanlis or Beurré Hardy, must first be used as an intermediary, the incompatible variety then being grafted onto this, when the result will be completely satisfactory. William's Bon Cretien, Dr. Jules Guyot, Packham's Triumph and Bristol Cross, require " double-working."

It should here be said that pear trees on quince stock should be planted with the graft at least 3 inches above soil level, so that scion rooting does not occur. This would give additional vigour at the expense of cropping.

Again, it should be remembered that pears on quince require a richer and more humus laden soil than those on pear stock, which will tolerate and may even crop better where the soil is neither so deep nor so rich.

CHAPTER V

POLLINATION AND FERTILITY

Pollination of Cox's Orange Pippin—Biennial bearing—Triploid varieties—Building up a collection of fruit trees—Flowering times—The Pear and pollination.

i. THE APPLE

WHEN PLANTING new fruit trees there is a great tendency to neglect the pollinating factor, with the result that though the stock may be of the best and have been planted with care, nothing but disappointment will be the result. Certain varieties are self-fertile and so will, up to a point, set their own fruit; some are partially self-fertile, and so will set a partial crop without a pollinator. These are the Diploids. Others, called Triploids are almost sterile and so must be given a suitable pollinator to help them set fruit. The question of providing a suitable pollinator is not easily worked out, and yet to obtain heavy crops it is essential. It is of little use, rushing to post off that order for that ' bargain offer ' of Cox's Orange Pippins, your favourite apple, in the hope that they will bear plenty of fruit without a suitable pollinator. They will not.

POLLINATION OF COX'S ORANGE PIPPIN

It was Messrs. Backhouse and Crane, of the John Innes Institute, who first discovered that Cox's Orange Pippin would not set its own pollen, a fact still not realised by many gardeners who continue to plant this variety by itself in the expectations of a good crop. Research has shown that even where the blossoms of certain varieties open at the same time as the Cox's Orange, it does not mean that a completely successful pollination will result, for it has been shown that the pollen of such varieties as Stirling Castle, Merton Worcester, Worcester Pearmain, James Grieve and Egremont Russet, ensures a setting of fruit more than twice as high as that of certain other varieties.

A most interesting photograph which appears in *The Apple*, by

Sir Daniel Hall and M. B. Crane, shows one half of a tree of Cox's Orange, which has been crossed with Sturmer Pippin, covered in fruit, whilst the other half, self-pollinated, set only two small fruits. Experiments carried out at the John Innes Institution by these two famous fruit experts resulted in the interesting information that of 11,949 flowers of Cox's Orange self-pollinated, only 92 flowers or 0.76% set any fruit. Yet crossed or pollinated with Egremont Russet 13.4% set fruit; with St. Edmund's Russet, 14.4%; and with Stirling Castle 15.5%. Where Ellison's Orange, Worcester Pearmain and Sturmer Pippin were used separately as a pollinator, the set was just 8.0%. The results also clearly show that not only is Cox's Orange a poor pollinator for most other varieties, but also where Cox's has been used as a parent. For instance, crossed with St. Everard, of which Cox's Orange was used as a parent, the percentage of fruit set showed only 3.3%, though St. Everard crossed with Beauty of Bath gave a setting of 10.8%. Of all the tests carried out at the John Innes Institute, the average fertility from cross-pollination showed that 11 blooms set fruit out of every 100, yet where varieties were self-pollinated the average of fruit set was only 2.5%.

Generally a fair test is to plant with the self and partially fertile varieties, another which will be in flower at the same period, and which should be planted as near as possible, so that the bees, and other insects, are able to transfer one lot of pollen to the other without having to travel long distances. For example, with Royal Jubilee, which blooms very late and so misses late frosts, plant the equally late flowering Crawley Beauty or Edward VII, or Lane's Prince Albert, which has a prolonged flowering season. It is interesting that where several trees of Royal Jubilee have been planted by themselves, they set only .09% of fruit, yet planted with Lane's Prince Albert the percentage was extremely high, as much as 16.1%. So do not blame the nurseryman, or your gardener, if your trees do not bear heavy crops! Gardeners of former times got over the trouble by planting dozens of varieties together, but then they had large gardens in which to do so.

BIENNIAL BEARING

Again, the question of biennial bearing must be considered. Ellison's Orange, a useful pollinator for Cox's tends to biennial cropping, and if this apple is used as a pollinator, another, possibly Worcester Pearmain, to take over on its 'off' season, should be planted with it. These apples also tend to biennial planting:

Allington Pippin, Bramley's Seedling, D'Arcy Spice, Miller's Seedling, Newton Wonder.

The choice too must be governed by climatic conditions, for James Grieve is not suitable for planting in the moist West of England, and so Worcester Pearmain may be used in that region, and James Grieve confined to the drier midlands, parts of Scotland, and the South East, should Cox's Orange be planted in these parts.

There is also the question of lime sulphur spraying to combat scab and mildew. Here certain varieties are sulphur shy, leaf and fruit drop often being the result. It will therefore be essential to plant together only those apples which are unharmed by lime-sulphur. St. Cecilia, Beauty of Bath, Newton Wonder and Lane's Prince Albert are all fairly sulphur shy, whilst Newton Wonder, a really grand quality apple, is also inclined to biennial bearing, so it would be advisable to plant several of these apples together. Or plant Newton Wonder with Lord Derby, which though tolerant of lime-sulphur is extremely resistant to scab and mildew, and no spraying may, in any case, be necessary. The two are also excellent pollinators.

TRIPLOID VARIETIES

In many years connected with orchards, I have found that three suitable varieties planted together will give the largest set of fruit. This is certainly true when planting biennial croppers and the triploid varieties, those which will not cross-pollinate with each other, and which are not very good pollinators for others. They must be planted with diploid varieties, which fortunately most apples are, and with those which bloom at the same time and are not given to biennial cropping. It is therefore wise to use two diploid pollinators, for there also may be the additional loss of the blossom of one variety through frost damage, and in addition the diploids will pollinate each other. The popular Bramley's Seedling, is a triploid variety, and should be planted with Grenadier, an early cooking apple; and with James Grieve or Lord Lambourne, for providing dessert. Those three fine dessert apples, Blenheim Orange, Gravenstein and Ribston Pippin, are all triploids and should never be planted together. With Ribston Pippin and Gravenstein, plant Lord Lambourne; and with Blenheim Orange, plant Egremont Russet, and they will give no trouble, for neither pollinator tends towards biennial bearing. Beauty of Bath is also a suitable pollinator for Ribston Pippin and Gravenstein, but is of course sulphur shy, which the others are not. To fertilise the pollinator itself, it is advisable to plant another similar flowering diploid as previously explained.

BUILDING UP THE COLLECTION

When commencing with apples most wish to begin with Cox's Orange, the best eater, and with Bramley's Seedling, renowned as the best cooker; then plant with them James Grieve which will pollinate both and add Grenadier, another good cooker, as a second line of defence, for this will also pollinate both. You will then have an early and a late apple for both cooking and for dessert. Then others can be added by degrees. Laxton's Advance, a grand early apple is also a Cox's pollinator, though not so good as James Grieve, but as it will also pollinate Bramley's Seedling, it may be a much better proposition in the wet districts. Then add Laxton's Superb, also a reasonably good Cox's pollinator, but which inclines to biennial cropping and so should be assisted with Laxton's Advance, or with Worcester Pearmain, or Fortune. And so on.

Of those that are very self-fertile, such as Laxton's Exquisite, and Epicure, Worcester Pearmain, St. Everard, Christmas Pearmain and Rev. Wilks, two only with the same flowering period need be planted together, and one would be assured of a satisfactory pollen setting, but even though each is self-fertile, it is inadvisable to plant on their own for only a proportion of the expected maximum crop will be the result.

FLOWERING TIMES

Flowering times are interesting for whereas Ribston Pippin and Wagener flower early in the season, they are classed as fairly late dessert apples. And in bloom at the same time, Gladstone and Beauty of Bath, are the first apples to mature, in the West Country being ready for use in late July. The average length of time for all apple trees to bloom is about 15 days, spread over a period of about 30 days, the very early flowering varieties being in bloom for the first 14/15 days or so, whilst the mid-season blooming varieties partially overlap the last few days of the early flowering varieties, and the first days of the late flowering varieties. The very late flowering apples, e.g. Edward VII, Royal Jubilee and Crawley Beauty, do not come into bloom until the last of the mid-season varieties have finished flowering, with the exception of the very long blooming, Lane's Prince Albert. In selecting suitable pollinators one must be governed by the flowering period of the trees and not by their maturity.

These apples bloom very early, depending upon climatic and seasonal conditions, and will pollinate each other:—

POLLINATION AND FERTILITY

T. = Triploid; P.S.F. = Partly Self-Fertile;
S.F. = Self-Fertile; S.S. = Self-Sterile.

Beauty of Bath	(P.S.F.)
Bismarck	(P.S.F.)
Gladstone	(S.F.)
Gravenstein	(T. S.S.)
Keswick Codlin	(P.S.F.)
Laxton's Advance	(P.S.F.)
Laxton's Exquisite	(P.S.F.)
Laxton's Fortune	(P.S.F.)
Lord Lambourne	(P.S.F.)
Miller's Seedling	(P.S.F.)
Rev. W. Wilks	(S.F.)
Ribston Pippin	(T. S.S.)
St. Edmund's Russet	(P.S.F.)
Wagener	(P.S.F.)

These apples bloom early mid-season:—

Allington Pippin	(P.S.F.)
Annie Elizabeth	(P.S.F.)
Arthur Turner	(P.S.F.)
Bramley's Seedling	(T. S.S.)
Cox's Orange Pippin	(P.S.F.)
Egremont Russet	(P.S.F.)
Ellison's Orange	(P.S.F.)
Grenadier	(P.S.F.)
James Grieve	(P.S.F.)
King of the Pippins	(P.S.F.)
Laxton's Epicure	(P.S.F.)
Mother	(P.S.F.)
Orlean's Reinette	(P.S.F.)
Peasgood's Nonsuch	(P.S.F.)
Stirling Castle	(P.S.F.)
Sturmer Pippin	(P.S.F.)
Tydeman's Early Worcester	(P.S.F.)
Worcester Pearmain	(P.S.F.)

These apples bloom late mid-season:—

Blenheim Orange	(T. S.S.)
Charles Ross	(P.S.F.)
Claygate Pearmain	(S.S.)
D'Arcy Spice	(S.S.)
Early Victoria	(P.S.F.)
Howgate Wonder	(P.S.F.)
Lady Sudeley	(P.S.F.)
Lane's Prince Albert	(P.S.F.)

Laxton's Superb	(P.S.F.)
Lord Derby	(P.S.F.)
Monarch	(P.S.F.)
Newton Wonder	(P.S.F.)
Rival	(P.S.F.)
Sunset	(S.F.)
Warner's King	(T. S.S.)
Winston	(S.F.)

These apples bloom very late:—

Crawley Beauty	(S.F.)
Court Pendu Plat	(P.S.F.)
Edward VII	(P.S.F.)
Royal Jubilee	(P.S.F.)

As a general rule varieties from each section should be planted together to obtain the best results from pollination, though many will overlap, such as Lane's Prince Albert, which has a very long flowering period, about 21 days, with Edward VII and Crawley Beauty.

ii. THE PEAR

As with apples, some pears are triploid varieties, and not only require a diploid pollinating variety flowering at the same time, but a second diploid variety which will be able to pollinate each other.

The following are triploids:—

Beurré d'Amanlis	(E)
Catillac	(L)
Jargonelle	(E)
Pitmaston Duchess	(L)

From this it is seen that two are early flowering, and two bloom late. With either then, it will be necessary to plant two early or late flowering diploids.

These pears bloom early:—

Beurré Easter	(S.S.)
Beurré Hardy	(S.S.)
Beurré Superfin	(S.F.)
Conference	(S.F.)
Durondeau	(S.F.)
Louise Bonne	(S.F.)

POLLINATION AND FERTILITY

These pears bloom mid-season:—

Beurré Bedford	(S.F.)
Clapp's Favourite	(S.S.)
Emile d'Heyst	(S.S.)
Glou Morceau	(S.S.)
Josephine Malines	(S.S.)
Thompson	(S.S.)
William's Bon Cretien	(S.F.)

These pears bloom late:—

Dr. Jules Guyot	(S.F.)
Doyenne du Comice	(S.S.)
Fertility	(S.S.)
Laxton's Superb	(S.F.)
Marie Louise	(S.F.)
Winter Nelis	(S.S.)

Strangely with pears, several varieties are unable to pollinate each other though both may be in bloom together. The very fertile diploid, Conference, is unable to pollinate the triploid Beurré d'Amanlis, though both are in bloom at exactly the same time. So do not plant them without another early flowering pollinator, such as Beurré Hardy, or Durondeau, and expect to obtain a heavy crop as is so often done, though both are hardy, easily grown varieties. Neither will Seckle pollinate Louise Bonne, though both are partially self-fertile and in bloom together, early to mid-season. All of which requires careful consideration before ordering your pear trees.

It should be said that though a number of varieties are self-fertile, and able to set their own pollen, they will bear a much heavier crop if planted with other varieties in bloom at the same period.

Early and mid-season flowering varieties, and mid-season and late flowering may be planted together, and may be relied upon to pollinate satisfactorily.

CHAPTER VI

PLANTING FRUIT TREES

Purchasing reliable trees—Soil conditions for planting—
Planting the trees—Staking and tying.

WHEREVER POSSIBLE see the fruit trees growing, which will reveal how they are grown and their quality, before placing your order. Even if no more than three or four trees are to be planted, remember that they will be planted with the expectation that they will bear good crops for at least a lifetime, and whereas to lose £2 on badly grown trees may prove disappointing, to have those same trees in one's garden for thirty years or more thereby depriving the small garden of producing home-grown fruit, would be even more disappointing. If possible, obtain your trees from a local nursery, or one situated as near to one's garden as can be found, that is provided the quality is above reproach. Trees which have been reared under similar conditions of soil and climate as to be found in one's own garden may be relied upon to give a good account of themselves when moved only a short distance.

Another point which should be given priority is that where a particular variety is required, then it is only common sense to obtain this from a nursery in the district where the variety was first raised, or first found growing, for a particular variety must evidently have proved vigorous where first observed e.g. Bramley's Seedling in Nottinghamshire; Newton Wonder in Derbyshire; Crawley Beauty in Sussex; Woolbrook Pippin in Devon.

Of course it may be that some of the less well known varieties may only be obtained from certain nurseries, and some firms may be specialists in the production of apples and pears, others in the stone fruits, all of which should be considered before purchasing the trees. And guard against those so-called 'bargain offers,' so often of stock which had been purchased on the continent very cheaply, and just heeled in upon arrival in this country. With the present high cost of labour and land, fruit trees cannot be produced cheaply, that is trees of top quality, and anything inferior will be money wasted. The very best trees will cost only a shilling or so more than those 'bargains,' yet they will if all other things are considered, provide a lifetime's pleasure, and occupy no more room in the garden than cheap, badly grown trees.

An important advantage of obtaining one's trees within reasonable distance of one's garden, is that they may be transplanted with as little root disturbance as possible, being moved with a good proportion of soil, whilst the roots are not exposed to strong winds more than necessary. Again it is possible either to collect or have the trees delivered just when they are required for planting. If the trees are to be planted during the December to February period, especially in the North, the trees may arrive from a nursery where hard frost is not being experienced, at a time when one's own garden is experiencing a long period of frost and snow. Should this happen, the trees must be placed in a shed or cellar, and sacking or straw should be placed over their roots until the soil is in right condition for planting.

SOIL CONDITIONS FOR PLANTING

Fruit trees are moved and planted between mid-November and the end of March, depending upon the condition of the soil, and this will be governed by its texture. A light, sandy soil will prove suitable for planting at almost any time during the winter months, for it never becomes sticky even following prolonged rain. But to plant in a heavy soil, however well prepared, will be to make it so compact as to cut off the vital supplies of oxygen to the roots, and the tree will be unable to get away to a good start, and so may never flourish as it should.

Where excessively wet soil conditions occur, the trees upon arrival should be placed into a 12 inch deep trough, the straw with which the roots are packed being placed over the roots, and over which the soil is replaced without in any way being made firm. Where planting cannot be done for business reasons, or for reasons of ill-health, the trees should, if possible, be given the same treatment upon arrival, for it is important to prevent the roots from drying out. Again this is where it is advisable to purchase from a firm with a reputation to uphold, for the trees will be sent out correctly packed in straw and sacking, and despatched either by 'goods' or 'passenger' service at the customer's request, the additional carriage generally being borne by the customer. For a small additional charge too, the nursery will name the individual trees with a metal name-plate, in place of the usual paper label which will quickly decay. Having the trees clearly and correctly named, adds interest to one's garden, and it will be a help with pruning.

It should be said that it is possible to move fruit trees as late as early April, if lifted with a good ball of soil about the roots and replanted almost at once. This does not mean that this is the most suitable time of course, for then the trees will be coming into activity after the period of winter dormancy, which also means that root action will have commenced. But it is possible that those taking over a new garden may not be ready to plant until early April, say at Easter (if not too late), when the business man generally does his planting.

Before making the holes to take the trees correct planting distances should be obtained, so that the trees are in no way deprived of air and sunlight. This may have been done before the trees were obtained, when the ground was being prepared so that the manures were used without undue waste.

Where planting cordons in a row, the best method is to make a trench for here the plants will be placed just over 2 ft. apart, wide enough to ensure that the roots do not touch when planted.

Those trees to be planted against a wall will require an additional amount of humus to conserve summer moisture, and decayed leaves, manure or moist peat may be worked in at planting time.

PLANTING THE TREES

The trenches or holes to take the plants should be made before the trees are taken from where they have been heeled in, or kept covered from frost, so that the roots are not unduly exposed to the air, and especially to a drying wind so often experienced in early spring. As a general rule trees for a light soil, and where of a chalky nature, should be planted during November to December. Those for a heavy, cold soil are best planted in March. Equally important is depth of planting.

Failure for the tree to bear well over a long period is so often due to either too shallow or too deep planting, and both contribute equally to the various causes of failure. Too shallow planting will cause the roots to dry out during a period of prolonged drought, and especially where the soil lies over a chalk subsoil. It may also cause the trees, where Type MIX rootstock is being used, to fall over even when fully established, and especially where planted in an exposed garden.

Too deep planting on the other hand will mean that the roots will be in the cold, less fertile subsoil, cut off from air and the sun's warmth, whilst it will mean that the scion, at which point the graft has been made on to the rootstock, will be buried and may take root. This will mean that the characteristics of the rootstock will play little part on the habit of the tree.

When buying and planting fruit trees always bear in mind that the roots are as important to the tree, more so in fact, than its shape, and for this reason the younger the tree the more readily is it transplanted. Where planting an orchard, however small, maiden trees, 1 year old, are not only less expensive, but are more readily established and may be trained and pruned to the requirements of the grower, rather than to those of the nurseryman. Remember that with all trees, the younger, they are the more readily will they transplant, though in this respect the exception is the pear, which will readily transplant up to 20 years of age. After the hole or trench has been made to the correct depth, so that the level of the soil will be at a point just above the top of the roots, as near as possible to the same level as the tree was planted at the nursery, a spadeful of a mixture of sand and peat, to encourage the formation of new fibrous roots as quickly as possible, should be spread about the hole. To enable the roots to be spread out correctly, a small mound of soil should be made at the bottom of the hole.

The old gardeners would place a flat stone on the top of this mound to prevent the formation of a tap root. This may still have its devotees, but the shortening of any large tap root with a sharp knife just before planting should be all that is necessary, at the same time removing any damaged roots, or shortening any unduly long roots. Here again, the experienced nurseryman whose reputation is built upon the success of the trees sent out, will see that the trees are lifted as carefully as possible, with the roots in no way damaged. All too often those ' bargain parcels ' arrive almost rootless and take years to become re-established.

The roots should be spread out so that each one is comfortable. All too often trees are planted in holes which are made far too small, with the result that they are bunched up, and compete with each other for nourishment. A tree badly planted can never prove satisfactory.

When the roots have been spread out, scatter more peat and sand about them, then commence to pack the soil around them. This is best done by holding the tree straight, or at the required angle in the case of cordons, which should be fastened to the wires before the soil is filled in. By pushing in the soil with the feet, a strong pair of boots being the best guarantee of correct planting the job may be performed by one person. As the soil is pushed into the hole it is trodden firmly about the roots so that there will be no air pockets, which would cause the roots to dry out. Tread the soil in little by little so that a thorough job is done, rather than fill up the hole and tread down afterwards. Where planting against a wall it is advisable to incorporate additional humus materials, as the soil is being placed in the hole to retain the maximum of moisture about the roots. The same may also be done where planting in a light, sandy soil. The planting of a few trees may be much more thoroughly done than where planting in an orchard, and the work should not in any way be hastily carried out. If there can be two people to do the planting so much the better.

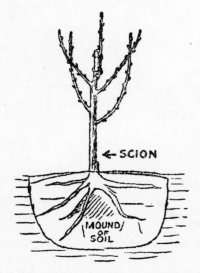

Planting a bush apple tree. The roots spread well out on a mound of soil.

STAKING AND TYING

Staking is one of the utmost importance. Bush trees, pyramids, and maidens, should require little or no staking, but cordons and espaliers should be fastened to the wires as soon as placed in position. Wall trees in the espalier and fan shape, must be fastened to the wall or trellis without delay, using a piece of strong leather for each shoot, which is correctly spaced to prevent overcrowding. It is the full and half standard forms which require careful staking, the stakes being driven into the soil so that there will be about 3 inches from the stem. If too near they will rub and cause serious damage to the bark. Care must also be taken to prevent similar damage to the base of the stems by boot or spade when planting.

Celia

Elton Beauty

TWO NEW DESSERT APPLES OF MERIT

Opalescent, an excellent dual-purpose apple

Exhibition plum, the new Thames Cross

PLANTING FRUIT TREES

It is advisable to use one of the patent tree fasteners rather than one which is home made, though an excellent use of old car inner-tubes may be made by cutting in 12 inch strips and twisting between tree and post to prevent rubbing.

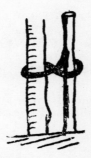

After planting and tying, give each tree, or row of cordons, a thick mulch of decayed straw manure, or even of decayed leaves to prevent loss of moisture whilst the trees are being established.

CHAPTER VII

TRAINING YOUNG TREES—
APPLES AND PEARS

Training bush and standard trees—Dwarf pyramids—Cordons—Espalier or Horizontal form.

BEING GROWN in exactly the same forms, apples and pears may be trained in the same way, either as:—
(a) Bush and Standards
(b) Dwarf Pyramids
(c) Cordons, or
(d) The Espalier or Horizontal form.

If your fruit trees are being purchased as maidens, that is 1 year old trees, which is not only the best, but the most inexpensive way, they will require training to form the required shape. And when once this form is accomplished, the trees will require careful pruning, not only to maintain their form, but to maintain a balance between growth of the tree and their cropping powers. This will mean the removal of surplus wood which would prevent the necessary sunshine and air from reaching the centre of the tree, not only necessary to ripen the fruit and prevent overcrowding, but also to ripen the new wood as it is formed, and on which the future crop will depend. To plant a top quality tree into well prepared soil, and then to allow that tree to grow away as it likes, will mean that quickly it will make an excess of wood, for the natural tendency of any fruit tree is to grow away as quickly as it can with very little thought to bearing fruit. Frequently quite the opposite is found in private gardens, the jobbing gardener cutting the tree so hard back each year as to deprive it of sufficient wood and foliage to enable it to bear a good crop. The result is that before long the tree is just a mass of old wood, the fruiting spurs become overcrowded, and gradually they decay with no new growth to take their place, and with the gradual diminishing of the crop. But before any serious pruning is necessary, the tree must first take form.

(a) TRAINING BUSH AND STANDARD TREES

This is the most popular form both for orchard or private garden planting, though for a small garden the more artificial forms are more suitable.

For a standard, what is known as a 'feathered' tree should be obtained. It should be two years old, and may be trained to any length of stem by removing the small 'feathers' or lateral shoots which appear on the main stem. The tree should be allowed to grow away without any check or pruning, then when the standard has reached the desired height, it should have its unwanted 'feathers' removed and the head is built up in the same way as for a bush tree.

The formation of the head will be either by one of two methods:—

(a) The Open Centre Form, or
(b) The Delayed Open Centre Form.

To form a bush on a good 'leg' by the Open Centre Form (a), and the head of a standard is formed in the same way, the maiden tree should be 'headed' back to about 3 ft. above ground level. This will persuade the tree to break into two or three shoots, which will become the framework of the tree, any shoots which break at the lower 15–18 inches of the stem being removed.

The following winter, which is the time when all pruning and cutting of fruit trees is done, the selected shoots are in their turn cut back half way, those with a more drooping habit, like Lane's Prince Albert being cut to an upward bud.

Next year these so-called extension shoots are likewise cut back about half way, or to about 9 inches of their base at which point they will have 'broken' or commenced to shoot. The head of the standard, or bush will now have been formed, and henceforth pruning is carried out for fruit bearing depending upon the habit of each variety, e.g. tip or spur bearer.

The Delayed Open Centre Form (b) is formed by removing only the very top 4–6 inches of the main stem. Then down the whole length buds are formed, and it is from these that the tree is built up. So as not to interfere with the laterals which will grow from the top two buds, the two immediately beneath should be removed. This will prevent the centre from becoming crowded. With this form it is the spacing that is all important, and to see that the shoots are facing in the right direction on all sides of the tree rather than too many appearing together. This is the best form for pear trees, preventing that 'drain-pipe' appearance of pears in the open centre form.

Open centre form.

(b) DWARF PYRAMIDS

This is a form of great value in the small garden, and which can be built up into a heavy cropping tree in as short a period as possible. As it is desired to make as much growth as possible at the beginning, and the tree to be brought into bearing early, bud growth must be stimulated. This is done by making a cut in the bark just above each of the buds on the main stem, taking care to select buds suitably spaced. These shoots may be pruned back to half the new season's growth each year so as to stimulate the formation of fruiting buds, and all blossom buds forming on the leader should also be removed. When once the tree comes into bearing it should be thinned out as for other forms by using one of the proven systems of pruning. Throughout its early life, and until thoroughly established, the main or central extension shoot must be constantly pruned back so that the tree can concentrate its energies on the formation of branches.

Dwarf pyramid form.

TRAINING YOUNG TREES—APPLES AND PEARS

(c) CORDONS

It is the single-stemmed cordon that is most frequently used and which should be planted at an oblique angle, so as to limit its tendency to grow away. Once again, a dwarfing rootstock should be used and neither a vigorous tree like the Bramley, nor a tip bearer like the Worcester Pearmain. Likewise the upright spur bearers of pears should only be used. Those apples and pears with a drooping habit do not make good cordons.

The maiden trees should be planted 3 ft. apart, and should be fastened to wires at an angle of 45°. The extension or main stem is never pruned, and in the early years pruning consists of cutting back the laterals during August to 6 inches from the main stem. This summer pruning will ensure the formation of fruiting spurs as quickly as possible. When the tree has made the necessary growth, the leader may be cut back so that the tree can concentrate on the formation of fruit rather than on extending its form. Henceforth, the tree may be kept healthy and the fruit of a high quality by the careful elimination of surplus spurs and a tree with excessive vigour may be curbed with root pruning done every three years. But by keeping the stem at an oblique angle this will also retard the formation of too much new wood. By planting apples and pears in the cordon form, this will allow a much greater variety to be enjoyed, whilst at the same time providing better pollination where space permits the planting of very few trees.

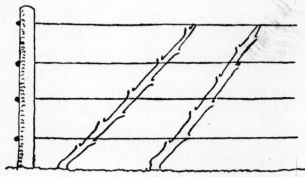

Besides the Single or Oblique cordon, the U-shaped form, or Double cordon should be understood as this is occasionally required. Though growing in an upwards direction as against the angle of the single cordon, the bend at the bottom will act as a check to vigorous growth. The U-cordon will be grown against a wire fence as in the case of espaliers and single cordons. Its formation is in fact very similar to that of the horizontal trained tree, the maiden being cut back to 12 inches of stem to two buds facing in the opposite direction. These are allowed to grow unpruned throughout the year being fastened to canes against the wires, first at an angle, then gradually to a vertical position.

Pruning consists of cutting back the leaders each autumn to one-third of their new season's growth, and of pinching out any side growth during August. These side shoots may be further cut back in November to two buds, which will form the fruiting spurs. A variety showing excessive vigour may be root pruned in alternate years. Should either of the buds fail to form an arm, notching or nicking immediately above will have the desired stimulating effect.

(d) ESPALIER OR HORIZONTAL FORM

Trained horizontally along the wires in a similar position as for cordons, there is no more satisfying way of growing apples and pears adaptable to this form. A maiden should always be planted, the stem being shortened to about 18 inches above soil level, and to a point where there are two buds close together, one on either side of the stem. It is a simple matter to train the tree, the laterals formed by the two buds being tied to the wires to the right and to the left, whilst the extension shoot is allowed to grow away unchecked, until sufficient growth has been made for it to be cut back to two more buds similarly placed and spaced about

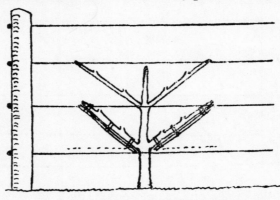

Training to the Horizontal form.

15–18 inches above the first to form. To encourage more rapid growth, the laterals should first be fastened at the angle of 45° and only placed in the horizontal position at the end of the first year's growth. Small canes should be used to train it at this angle otherwise there will be fear of damage by strong winds.

A new tier may be formed each season and when the first has been formed, to encourage it to form fruiting spurs, all shoots formed on these branches should be pruned back in summer to within 5 inches of the main stem. This will encourage the plant to form fruiting buds instead of new wood. The work should be done towards the end of July. This is followed by cutting back still further during winter in the usual way. Treatment then consists of thinning out the established spurs, and root pruning if the tree is inclined to make excessive growth. As the side arms continue to make wood, this new wood should be shortened back to a half of the newly formed wood each winter, again making certain to cut to a bud which is to form the extension shoot. This may continue for a number of years and until the branches reach the required number. To make for ease in picking and pruning, it is general to allow five pairs of arms or tiers to form, the top arms at a height of about 6 ft.

CHAPTER VIII

RENOVATING AND PRUNING ESTABLISHED TREES

The use and care of tools—Making a correct cut—The functions of Pruning—Renovating old trees—Pruning young trees—Restricting the vigorous tree—Root pruning—Bark pruning—Pears—Grafting.

APPLES AND PEARS

No PRUNING can be done without the aid of a good tool. Rusty and blunt secateurs will not only make what should be an interesting occupation, a most unpleasant job, being severe on the hands, but the points at which the wood is removed will be torn or bruised, providing an entry for disease. Pruning top fruit trees demands a more reliable tool than one used for pruning black currants, or gooseberries, and only the best make of secateurs should be obtained. The same pruners may also be used for rose trees and other shrubs, but what may be suitable for those plants may not be sufficiently strong for pruning fruit trees. A lady will perhaps select smaller pruners than those used by a man, but the same care should be taken in selecting a tool which will feel comfortable to use. Remember that to obtain heavy and consistent crops, regular attention to pruning is essential, and good pruning cannot be done without reliable secateurs.

Of those obtainable the Pruneesy Secateurs with its movable jaws, so that any sized wood can be cut with ease are an excellent make. Likewise Wilkinson's Pocket Sword Pruners, which open and close by a concealed spring, and are capable of cutting the most stubborn wood with surgical precision. Then there is the Rolcut Superlight model, with its press-button spring, most excellent and easily handled secateurs. Whichever one's choice, always take as much pride in and care of the tool as the mechanic does with his spanners, or the cricketer with his bats.

A pair of good pruners will be the fruit enthusiast's greatest friend, so when they have been used, wipe them clean with an oiled rag and replace them in their box,. My own pruners are used almost every day from November until January and every evening are oiled and replaced in their box, kept in a special drawer in my study. They are never placed in the garden shed to be picked up and used for cutting wire by anyone who happens to require them for this or any other purpose.

To reach the top branches of what may be old or neglected trees, a long-arm will be necessary. This is the nurseryman's term for pruners fastened at the end of a 6 ft. pole, and manipulated by a lever and stout wire. This tool will enable the pruner to reach to a height of almost 12 ft. from the ground, without the aid of a ladder or steps, which are often dangerous for an older person to use, though they may be required for picking the crop. The same care should be taken when using this tool as for the ordinary secateurs, first place the cutter blade against the wood before using the lever to sever it. Any undue tugging will only cause tearing or bruising, so often the cause of disease. The same when using hand pruners, never drag at the branch. If it is too tough for the secateurs, then make a clean cut with a small, sharp saw, kept oiled and free from rust with the same care as that given to the long-arm and secateurs. The oiling will not only keep the cutting blades free from rust, but it will also keep them sharp, and sharp tools are essential to good pruning and a healthy treee.

An efficient pruning knife is also essential for root pruning, notching, bark ringing and for the removal of suckers, which so frequently appear around the roots of plum trees. With its particular shape only a pruning knife will prove suitable for each of these operations.

MAKING A CORRECT CUT

It is important to know just how to make a cut before the

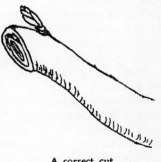

A correct cut.

pruners are taken up. This should be made immediately above a vigorous outward bud, and should be made with the slope of the cut away from the bud. The remaining wood above the cut will almost certainly die back to the bud, and if more wood than necessary is allowed to remain, this may become a source of infection. The cut should be made clean, without bruising or tearing.

It is also important to have handy a preparation for painting over the wounds caused by paring away diseased tissue and where branches have been severed. White lead paint is satisfactory, and a new liquid fungicide called Medo will do the job even more efficiently.

As soon as a tree has been pruned, the prunings must be cleared up and burnt. To leave them lying about the orchard or garden will only encourage disease, which may be eventually transmitted to the healthy plants from which they were taken. It is preferable to make a small fire towards the end of each pruning day and to burn the rubbish whilst still fairly dry and before the shoots become trodden into the ground.

THE FUNCTIONS OF PRUNING

Of all jobs in the garden nothing causes so much controversy, or so much worry as pruning. It would appear that, to the amateur fruit grower,, the arts and mysteries of pruning are only to be understood by those who have spent several years at a horticultural college, or as an employee at one of the fruit growing nurseries.

To begin with there is no doubt but that most gardeners carry pruning to excess, in fact to ridiculous limits, cutting back to the old wood so completely that the trees soon become devoid of any new wood at all. Pruning is done

    (a) So that a proper balance is obtained between leaf growth and root action,

    (b) To enable the requisite air and sunshine to reach the greatest possible portions of the tree,

    (c) To maintain a balance between the fruiting of the trees and the formation of new wood.

Pruning is not done for pruning's sake, or merely to try out the new secateurs received as a Christmas or birthday present. The same advice may be given the gardener as that given the author by the Yorkshire and England cricketer, Percy Holmes, when facing Leicestershire spin bowler, Jack Walsh, for the first time, "if in doubt, leave it alone." The advice was very successful even if the pads did more than the bat, and the same may be said of pruning, "if in doubt, leave it alone." Plums and cherries for instance, do not require anything like so much pruning as apples and pears, chiefly because of their habit of 'bleeding.' This will not only sap their strength, but will make them liable to disease entering through the wound. Instead of pruning the branches, the roots should be pruned. This will slow down top growth and the same result will be achieved without the risk of bleeding. Check the roots and check top growth and *vice versa*.

What is most important is to ensure the correct balance of the tree, between tree growth and root action, between the formation of fruiting spurs and the making of new wood. If this is maintained over the years, there is no reason why a tree should not remain healthy and bear a heavy crop for a hundred years or more.

Pruning is also done to stimulate growth, to make the necessary new wood which in turn will stimulate root action, and enable the tree to search vigorously for its nutriment. This means that contrary to general belief it is the slow growing varieties which require the most pruning in comparison to those of vigorous habit like Bramley's Seedling apple. To prune the vigorous grower hard back will be to encourage even greater vigour and the balance of the tree, the affinity between root action and top growth, may be destroyed.

FUNCTIONS OF ROOTS AND FOLIAGE

Continually it must be borne in mind that there is close affinity between root growth and foliage. The roots supply the raw materials that are to be turned to good account by the foliage which in turn supplies other substances which are continually building up a more vigorous rooting system, which in their turn are searching for other nourishment for the foliage to convert. By controlling either roots or foliage, this cycle is halted for a time, but it can be seen that to drastic pruning of either roots or foliage will have its effect in thowing the functions of the plant quite out of tune with one another. It is therefore necessary before doing any pruning to remember the close afinity between foliage and roots. If pruning is done at the wrong time of the year, it will cause severe harm to the functions of the plant. If the trees should be pruned whilst the leaves are fulfilling their functions in summer, the constitution of the tree will suffer; the roots will be searching out for nourishment which cannot be converted by the foliage. The result would be that soon the plant would die back altogether. That is the reason why fruit trees are pruned in winter, preferably between November and early March. A little judicious thinning to allow more light to enter may be in order, or pinching back of surplus shoots of plums and other fruits, but any severe removal of foliage during the period the sap is most active will mean a tree of reduced vigour, rather than one of additional vigour if pruning is done whilst the sap is dormant. Throughout the life of a tree the aim must be kept constantly in mind, that it is required first to build up a vigorous tree; to ensure a bloom of fruit of top quality, and then to maintain a constant affinity between health, vigour and quality, all of which is not difficult if we realise just why it is necessary to prune before we take up the pruners.

Pruning is also done to enable air and sunlight to reach all parts of the tree so that new growth is stimulated, and which will gradually replace the old wood. Where air and light is cut off from the tree, a preponderance of old wood forms, which will gradually die back and leave a sparsely furnished tree, one yielding only a small percentage of the fruit that it is able to do.

RENOVATING OLD TREES

It often happens that when moving into a new house, one may also take over a badly neglected orchard, or even one or two trees which have never been correctly pruned, yet which with some attention will bear good crops.

A ladder and a saw may be necessary to reach the top branches. If the trees are so badly overcrowded as to restrict air and light from reaching each, then do not hesitate to cut out several large branches, or even to remove a tree in its entirety. It should however be said that an old tree may die back if too many large limbs are removed at once. The same with trees as with humans, an old man may succumb to the loss of a leg, whereas a young person would be able to overcome the shock. It is therefore always preferable to prevent the necessity of removing large branches and this is only done by regular attention to the tree's requirements.

The first operation in renovating an old tree will be to cut out all dead and decayed wood which is playing no part in the life of the tree and which it will be better without, for the greatest source of disease will then be out of the way. Then look at each tree again and where one branch is possibly growing into another tree causing little light to reach it and obstructing the flow of air around each tree, cut this away also. When removing any wood, whether decayed or green, make the severance right against the main trunk, and just in the same way if decayed or surplus wood is to be removed from a small branch. Frequently it is observed that a branch has been removed an inch or even several inches away from the main trunk or branch, with the result that the

RENOVATING AND PRUNING ESTABLISHED TREES

remaining wood gradually decays and falls a victim to pest and disease, especially Brown Rot Disease, which will attack the remaining parts of the tree and also the fruit.

When removing a large branch it is advisable to give it some support whilst the cut is being made to take off much of the weight and so prevent the branch from tearing away from the stem and which would cause considerable damage to the bark.

It will be found that a cut made close to the bark will quite quickly heal over and so will be closed against disease. But it is often noticed with an old orchard that certain branches will have been carelessly removed, or may have snapped off leaving several inches of wood which will have decayed and come away leaving an unhealthy looking cavity on the main trunk, or on a large branch.

To prevent further decay, this cavity should be filled up with cement or if only a small opening, with putty, but if left untouched there will be the chance that disease may damage the tree past repair.

Another method of pruning known as de-horning may also be practised, and is a better method where growth has become dense at the centre, and where the removal of a complete branch would not bring about the necessary results.

It is generally the top branches which are de-horned, which means cutting them back several feet, possibly reducing them by at least half their length. A sloping cut should be made exactly as when making a cut with the pruners, so that moisture drains away, and the wound should be treated as described where a whole

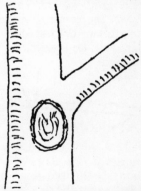

A branch carefully removed. New growth is beginning to cover the cut.

branch is removed from the main trunk. An abundance of fresh new growth will result and a completely new head may be formed.

As mentioned, any pruning and de-branching of an old tree must be done by degrees. The first winter, possibly no more than decayed wood and a few small branches overlapping each other will be removed. The following winter more unwanted wood may be cut away, then later if the tree has become excessively tall and straggling, it may be advisable to cut back the main branches to a sturdy young shoot, and so build up once again the lower part of the tree so that it will in time be capable of bearing a heavy crop. But any rejuvenation of an old, neglected tree must be done gradually. If you take out the saw and pruners and cut away right and left during the first winter there may be nothing but dying trees left. When a tree has been allowed to fall into neglect, the temptation to restore its vigour at once, is great, but it must be resisted. It may take four years to renovate an old tree, may be longer if the trees are very old.

PRUNING YOUNG TREES

When pruning neglected young trees, large branches will not need to be cut away. Instead, thinning and cutting back laterals to form vigorous buds will be all that is necessary. First remove any overlapping wood, taking care to cut back to an outward bud, for the centre of the tree must be kept as open as possible to let in the maximum of light and air. Then take a careful look at the laterals, which are the shoots growing out from the main stems and on which the fruiting buds are formed. Each season, additional wood is formed and also buds, but if not kept pruned the laterals will become longer and longer, and at the same time the buds will become weaker and weaker. Instead of allowing them to remain unchecked with the result that the fruit will be small, they should be cut back to two or three fruiting buds. Into these the energies of the plant will be diverted with the result that the fruit will develop to a good size.

Building strong fruiting buds.

Cutting back the unpruned laterals to two or three buds should be done before the buds begin to swell, in other words before mid-March, in order that when the sap commences to flow it can be directed at once to the fruiting buds and also there will be a danger of knocking off the buds if pruning is done when they have started to swell.

Varieties possessing extreme vigour, such as Bramley's Seedling and Newton Wonder, would be well able to develop four or five buds and too drastic pruning will only increase the vigour of the tree to the detriment of fruit.

As a general rule when taking over an established orchard, all outward and downward branches should be left untouched, unless overcrowding each other, for they will receive all the air and sunlight necessary. They also bear the heaviest amount of matured fruit, not only for this reason, but because they are better able to contend with strong winds, having much freer movement than those branches at the centre of the tree, with the result that the fruit is not blown off. It is the centre of the tree which requires the most attention for this should be kept as open as possible.

RESTRICTING THE VIGOROUS TREE

The very strong growing varieties such as the Worcester Pearmain and Blenheim Orange, together with those previously mentioned, will require very little pruning, for they do not need any stimulation to make fresh growth. Their growth may of course be regulated by planting a known rootstock of dwarfing habit, but this, of course, is known only where a new tree is being planted. But to prune the vigorous varieties without knowing which they are will cause only disappointment by increasing their wood to the detriment of fruit. Thus it is my contention that no orchard should be touched except to cut away decayed wood without first seeing the trees in fruit.

As it has been seen that to prune hard a vigorous growing tree, will only make it more vigorous, an overcrowded tree of this nature should have a branch or two completely cut away. This will allow the extra light to reach the buds without increasing its vigour. Or it may be restricted by either root or bark pruning. As a rule it may be said that a strong growing tree will form very many less fruit buds than will a slow growing tree, and so with the vigorous growers some method of restricting growth will often be necessary.

ROOT PRUNING

November is the best time to root prune, it being an easy matter to make a trench 3–4 ft. away from the trunk, and to sever the strongest roots, spreading out the more fibrous roots before filling in the trench. The same rule of careful pruning, doing only a little at a time apertains equally to the roots as it does to the branches, particularly where old trees are concerned. It is therefore advisable to root prune only one side of the first year, the other side the following year. Where standard trees are being grown it is not advisable to remove the tap root which is the tree's anchor. If a stone is used at planting time, the tree will concentrate on strong secondary roots which may if necessary be restricted by pruning. Nor should severe root pruning be done where the dwarfing rootstocks are used, nor should it be necessary, for their poor anchorage will be made even less secure.

It should also be remembered that root pruning should be consistent with the removal of wood to retain the balance of the functions of both roots and foliage. In the case of vigorous growers, the removal of a branch, or of unwanted wood should correspond to the restriction of roots. In dealing with old wall trees which are being root pruned in order to bring them into full bearing once again, the general practice is to prune back the fruiting spurs at the same time as the roots are cut back and this will ensure quality rather than a quantity of fruits of little value. Thus will the connection between roots and foliage be maintained.

BARK PRUNING

Bark pruning or ringing is done to curb the flow of sap with the result that more fruiting buds are formed instead of wood growth. As there is danger that too much bark may be cut away which would not heal over in a reasonable time, ringing should only be done when root pruning has no effect, but it is worth trying with a tree which refuses to bear a crop and is continually making fresh wood even when every known method of restriction has been tried. Instead of making a complete circle round the stem it is safer to make two half circles, allowing 6 inches of bark between each. Cutting should be done with a sharp knife; a pruning knife is best, and immediately the cuts have been made and about three-quarters of an inch of bark has been removed, tape should be fastened and bound securely round the place where the cuts were made. Early May is the best time to do this, and choose a calm day so that the tissue of the tree at the exposed place is not open to drying winds. Cover with tape immediately each tree has been treated.

PEARS

Everything that has been said about apple trees is much the same for pears, but here again we discover the name of each tree, and learn something of its habits before taking up the pruners. Pears are divided into two sections, those with a vigorous upright habit, and those of a weaker and semi-weeping habit. In the former group are Comice and Durondeau and Clapp's Favourite;

RENOVATING AND PRUNING ESTABLISHED TREES

of those with slender habit are Louise Bonne and Beurré d'Amanlis. The importance is in pruning for the upright growers to have their buds facing outwards, whilst the slender, weeping growers should be pruned so that the buds, as far as possible face in an upward direction. Most of the weepers are tip bearers and should be pruned but little for they make only a few fruiting buds, but those of vigorous habit may need to have their spurs reduced to obtain fruit of size and quality. The same remarks of the tip bearers in pears also concerns the tip bearers of apples, e.g., Worcester Pearmain, St. Edmund's Russet and Grenadier, and these trees will require but little pruning. But every variety should be treated on its merits. Do not over-prune any tree, first try the lightest possible pruning, then wait for the results. Never prune

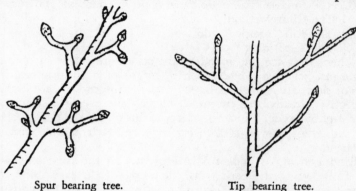

Spur bearing tree.   Tip bearing tree.

for pruning sake, and a little at a time is far better, especially with established trees, than being too drastic. First look at the trees then try to imagine them in fruit and remember that the aim is a healthy, well balanced tree, one able to bear the maximum amount of the best quality fruit and over as long a period of years as possible.

It should be noticed that each shoot or lateral will form both fruiting and wood or foliage buds, the former being easily distinguished by their habit of appearing on short, woody stems, whilst the wood-making buds lie flat along the stem, are smaller and of a pointed nature.

GRAFTING

This is an interesting occupation for the amateur who has taken over an old orchard, or where several old trees of a certain variety have not proved a success. Provided the condition of these trees is good and they are free from canker and other troubles, then they may be made to change their variety by top-working and frame-working. Top-working is done by removing the lower branches of the tree, and grafting buds of the desired new variety onto the topmost branches. Frame-working is done by allowing the branches to remain, and on to these is grafted the new wood.

Grafting is an art that comes only by constant practice. The work is done between early March and early June; the earlier in the season the better for satisfactory results, but the scions or prunings should be removed from the trees during January, for it is necessary to have this new wood in a dormant state at the time the grafting is done. The shoots should be tied up in bundles, named, and allowed to stand in a mixture of moist peat and sand under an open shed and given protection from severe weather until ready for use. There are three main methods of grafting.

(a) The Cleft Method;
(b) The Rind Method, and
(c) Side Grafting.

When ready for the operation the scions are cut at an oblique angle just below two buds, then by means of a grafting tool the cuts or clefts are made two, three or four at the end of each branch which has already been cut and trimmed. The prepared scions are then tightly wedged into the clefts, a little moist clay is rubbed in, and grafting wax is then poured over each at the point of insertion.

Rind grafting is done by cutting through the bark only and carefully lifting the bark from the wood of the tree. May and June are the best months for rind grafting and this is the grafting method most commonly used for apples and pears. The scion, prepared in the same way as for cleft grafting, should be inserted between the bark and wood at the point where the cut has been made. The jointing is sealed with wax and raffia, string or tape used to hold the scion to the tree, leaving the basal bud exposed. The process of 'marrying' soon takes place and constant watch is necessary, for as soon as the basal bud commence to swell the string or raffia must be cut. So much for the two main methods used in top-working.

For frame-working, side grafting is most commonly employed. The lateral shoots are first removed and the tree then worked with side grafts. A cut is made at an angle of about 25° on the selected branches, and into this is inserted the prepared scion which has received a cut at a slightly less oblique angle than for the methods previously described. Oblique side grafting is performed by making the cut on the branch at an angle of about 40°, and the scion is inserted in the same way. Grafting wax is poured round the cut and the operation is complete; and nothing could be more simple.

There are numerous other methods of grafting and these will be found in Mr. R. J. Garner's *The Grafters' Handbook*. For ordinary purposes those I have described should be sufficient. There is one more precaution to take, and that is to provide the grafts with some protection from insect attacks; for these may cause considerable damage by eating the scion buds. Grease bands tied round the trunk below the lowest branch will prevent certain species of weevil from crawling up, and this should be a routine precaution with all fruit trees.

It is of course possible to graft several varieties onto the same tree as done in the production of *The Family Tree* sold by R. C. Notcutt Ltd., of Woodbridge. The question of cross-pollination and of the most suitable varieties for certain districts, such as late flowering varieties for gardens troubled by late frosts, must all come into the reckoning. Rootstocks, too, are given consideration where new trees are being grafted with numerous varieties.

Where one's garden is especially small, this method of enjoying home-grown fruit has much to recommend it, for trees may be supplied in the bush or standard form, and crop remarkably well. Those grafted with especially compact varieties may also be used for planting in tubs for a city veranda.

CHAPTER IX

CARE OF YOUNG TREES

Building the framework—The established spur system—The regulated system—The renewal system—Summer Pruning—Biennial cropping—Branch bending—Notching and nicking.

APPLES AND PEARS

IN THE correct treatment and care of a young fruit tree lies its ultimate cropping powers, which include its health, vigour, shape and ability to bear a heavy crop of quality fruit as soon as possible and over as long a period as possible. There is a wide choice of types of tree available; the bush form, standards, cordon, fan-shaped and horizontal trained, and each demands rather different treatment not only in its establishment, but in its subsequent care when established. The interesting question of training can be left until the following Chapter, but here we are concerned with the care of the young tree after it has been formed.

First it must be remembered that in its early years, a young tree should not be expected to bear excess fruit at the expense of making a healthy frame; at the beginning, the formation of wood is more important than fruit, for a solid and lasting foundation must be formed.

Mr. George Bunyard of Kent always advocated that a newly-planted tree, which would be between 2–3 years old, should be allowed to grow away for a full season entirely untouched. This is to allow it to form ample new wood, whilst the new roots are forming, and so the balance of the tree is left undisturbed. There is then no fear of excessive pruning interfering with the functions of the rooting system whilst setting in, or whilst the head or form is being acquired, little or no pruning should be done.

The following winter, pruning may commence and one should have then formed an idea as to the system to follow. It will be one of three alternatives:—

CARE OF YOUNG TREES

(a) The Established Spur system, generally carried out for the more artificially trained trees of apples and pears.
(b) The Regulated system which requires the minimum of pruning and which is generally carried out on trees with a vigorous habit, and
(c) The Renewal system, which simplified, means keeping the tree in continuous growth.

THE ESTABLISHED SPUR SYSTEM

The value of this system is to allow the tree a greater freedom of growth with the formation of fruiting spurs along the main branches. Wood formed during the summer is cut back during winter to four buds. During the following summer the two top buds will make new growth, whilst the lower spurs will develop into fruiting buds. From the place above the top buds where the cut has been made, two laterals will have formed during the second season, which in turn are cut back (B) to two buds. Thus after two seasons you have this:—

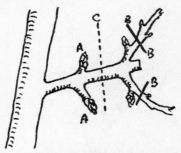

Forming a fruit spur.

This method will ensure that whilst the tree is concentrating its energies to the formation of fruiting spurs at (A), the balance of the tree is being maintained with the spurs forming fruiting buds without having to form new growth themselves.

During the third winter, the fruiting spurs being now correctly formed, the previous year's wood is cut back at (C), for its functions are now complete and the energies of the tree can concentrate on the production of fruit at the two spurs (A). Again to encourage the building up of a strong fruiting spur, the laterals should be pinched back during mid-summer, reducing them by about a third. In this way a tree is built up to its full fruit-bearing capacities in the quickest possible time, bearing in mind the affinity between the rooting system and the formation of foliage, both necessary to maintain the vigour and health of the tree.

PLANTING FRUIT TREES

When once the tree is established, little pruning will be necessary other than to remove any overcrowded branches and to cut out all overcrowded spurs. This method is of course suitable only for the spur-bearing varieties such as Cox's Orange Pippin, Christmas Pearmain, etc., and it is the trained forms which generally respond to this method the best. The method of thinning the fruiting spurs has been described in the previous Chapter.

A word should be said about the necessity to thin out the spurs when once the trees have become established, when they are about ten years old, and those taking over a garden with trees of this age should remember that if no spur thinning is done the tree may soon exhaust itself by forming excess fruit, too much for it to carry in comparison with tree growth. Most gardeners are shy at removing fruiting buds, but in any case too many will cause a reduction in the size and quality of the fruit and especially where this is the habit of the variety, e.g., Winston.

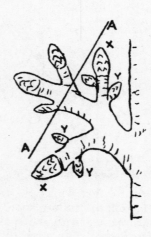

Treatment of an established spur to encourage large fruit. Remove at A.
(X = fruit bud; Y = foliage bud)

The question of the tip bearers does not come into this system for the buds are formed differently, and in any case they are not suitable for training in the artificial forms.

THE REGULATED SYSTEM

This system is more suitable for the tip bearers and for bush and standard trees of all apples and pears, for these are the most natural form of fruiting trees. By using vigorous rootstocks like Malling II for apples and Quince A for pears, the tree will not come into bearing as quickly as if the dwarfing rootstocks are used, but it will retain its vigour and its fruiting capacities over a longer period. With the tip bearers any excessive pruning will cause greatly diminished cropping, for the buds are borne at the end of the laterals and not in clusters as with the spur bearers. So under this system cut away as much over-crossing and centre wood as to keep the tree ' open ' and also remove all in-growing laterals as they are observed each season. Any strongly growing branches which appear to be growing away too quickly should be cut back, or de-horned as it is called, to a lateral growing out in a manner that will encourage the shape of the tree. The spur bearers should have their spurs thinned out in the way previously described and though this will not be so essential as with those trees growing on a dwarf rootstock or in artificial forms, overcrowded spurs should be regularly thinned. This system demands just as constant attention in the pruning programme and a little cutting back should be done each year, rather than the removal of excessive wood in alternate years. Those varieties which are excessively strong growing should be root-pruned every three or four years, for if too much de-horning is done this may only increase the vigour and too much wood will be the result to the detriment of fruit.

When forming new fruit trees, the less pruning the better until the fruiting buds have started to form, for pruning tends to encourage excess wood at the expense of fruiting buds. So first allow the trees a full year's growth before pruning, then commence with the bush and standard forms by shortening back the new season's wood of the leaders by a third at the end of every season. Then as the tree begins to take shape the leaders will require only tipping each year and possibly occasional de-horning if growth is too vigorous. The laterals too, for spur bearers will require cutting back as described, the tip bearers being left untouched until the time when they become excessive and some wood may have to be cut away.

THE RENEWAL SYSTEM

This works out exactly as described; it is the continued renewal of old wood by new, thus retaining the vigour of the tree over a very long period without making too much old wood. The idea is to maintain a balance between the production of new wood and fruit buds, and so it is necessary to build up an open tree with well spaced, erect branches for it is on these that the new wood is continually formed. Suitable erect growing shoots or leaders which form on these branches are pruned back to form replacement branches which in due course will take the place of the older branches. The same method takes place with side shoots. These are left to fruit unpruned. They are then cut back to within two buds of the base which will then produce two more shoots. Again, these are left unpruned and allowed to fruit. In turn, each of the two shoots are pruned back to two buds after fruiting and so the process of the continued replacement of new wood for old goes on. The proportion of shoots pruned and left untouched will be governed by the vigour of the variety. For Bramley's Seedling a better balance will be maintained if a greater proportion of shoots are left unpruned, for stimulation is not required. But much will depend on the general health and vigour of the variety or tree. If the tree seems to be making heavy weather of life, it will require more pruning of side growths to provide the necessary stimulation. Where a tree is healthy and vigorous a large number of the shoots may be left unpruned for as long as three or four years, thus maintaining a balance between fruit and wood.

SUMMER PRUNING

Those who grow apples and pears in the artificial forms, especially as cordons and espaliers, may prevent the formation of too much new wood by summer pruning. In the case of bush and standard trees planted in an orchard, pruning is done only during winter to stimulate the formation of new growth. By careful attention to the shortening back of wood during summer, this will provide a check on the action of the plant's roots, which in turn will check the production of new wood.

Mid-July is the time to pinch back any lateral shoots which have made excessive growth, and these should be shortened to 6 inches of their base. This should be done before the wood has ripened and before the end of the month!

Not only will the effect be to restrict vigour, but it will enable new fruiting buds to be formed more quickly than where summer pruning is not carried out, the tree using its energy in this way rather than in the continued formation of new wood.

There is another factor worth considering, where summer pruning is done, and that is to enable additional sunlight to reach the swelling fruit, and especially with pears, this will ensure more satisfactory ripening of the fruits.

BIENNIAL CROPPING

With apples there is a tendency for certain varieties, e.g., Blenheim Orange, Miller's Seedling and Newton Wonder, to form an extremely heavy crop only in alternate seasons and some form of regulated pruning is necessary to limit the blossom during the 'on' year, so that the fruit will retain its size and to encourage the tree to bear a certain amount the following, or 'off' year. All spurs should be reduced to two buds and all maiden wood must be left unpruned. This will ensure that blossom buds will be numerous for the following season whilst during the 'on' year, much of the vigour of the tree will be utilised in the formation of new buds rather than on the concentration of those already formed. Where possibly severe frost has damaged the blossom for one year it frequently happens that the tree will produce an excessive amount of fruit the following year and this may be followed by a lean year. By employing these methods a more regular crop will be assured.

A variety required for exhibition expected to make a good-sized fruit very early in the season, such as Arthur Turner and Emneth Early apples; and pear Doyenne d'Eté, it will be necessary to cut back their laterals more than is normally done. Instead of cutting back a third of the wood, as much as two-thirds will ensure a more rapid maturing of the fruit even if the quantity may not be so high as when given normal treatment.

BRANCH BENDING

Where a tree is making such excessive growth that there is a fear that further pruning will only stimulate more wood, branch or shoot bending will help to restrain this and at the same time by restricting the flow of sap will help to form fruit buds at the expense of new wood. It is the lower branches which are more easily bent and they may be either tied at their tips to the stem of the tree or may be weighed down with strong stones or bricks fastened with cord to their tips. Whilst this does not really come under the title of pruning it does restrict growth of fresh wood and will also cause these bent branches to bear a heavy crop of fruit. In this way, old fruiting shoots can be constantly replaced by new wood.

NOTCHING AND NICKING

It often happens that a certain bud is required to be restricted in its growth whilst another may refuse to break and so may cause the tree to become mis-shapen. To encourage a bud to break, a notch should be made above it, and to retard a bud, a notch should be made on the stem immediately beneath it removing a small piece of bark.

In general it is found that the most vigorous buds are those towards the top of a stem or branch, the vigour diminishing with the buds at the centre and being less vigorous at the lowest point. It may therefore be necessary to stimulate those at the lowest point by notching and this would even out the formation of new branches.

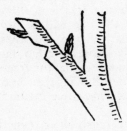

Nicking of the stem has a similar effect. It is generally done where restriction of the extension of lateral growth is required. In addition to the cutting back of the stem, the prevention of an extension by the top bud will ensure that those buds lower down the branch will make more growth which may be the object when building up a bush or standard tree. The cut or nick should be made with a sharp knife, preferably the pruning knife.

CHAPTER X

HARVESTING AND STORING FRUIT

The interest of a fruit store room—Apples—When to harvest—Factors denoting maturity—Storing conditions—Pre-harvest drop—Pears—Harvesting and Storing—Plums—Cherries.

NOTHING IS more interesting nor so valuable as a store well filled with long keeping apples, to a large family as valuable as potatoes, and it is always our aim to start the winter with a good stock of apples, potatoes and logs, for come what may we can keep warm and will not go short of food. But for the connoisseur of a dessert apple the store room is the most interesting part of the house or garden, bringing a breath of the summer during the long winter days. And how interesting are the old apples, to walk round the shelves of a well stocked store room is to turn back the pages of history, Gravenstein reaching this country from Germany two years after the Battle of Waterloo, whilst Ribston Pippin was first planted in the grounds of Ribston Hall, Knaresborough, during the year of the Union with Scotland, with Queen Anne still on the throne.

APPLES
WHEN TO HARVEST

One great mistake is to remove apples from the trees far too early. If allowed to hang as long as possible, the fruit will not only be reaching full maturity without which it will not keep, but will be far happier on the tree than in the best of store rooms, and its keeping qualities will be both lengthened and enhanced. Several of the russets, particularly Brownlee's Russet will shrivel and lose their crisp flavour if picked too soon. And Sturmer Pippin, a grand apple for a heavy clay soil and which will keep well into May, must be fully ripened or it will keep no longer than February.

We tend to harvest our late maturing apples during the

favourable weather of early October, for it can be an unpleasant business if the fruit is allowed to hang until the cold days of early winter, but many of the late keeping varieties are best left until then. One of the most delicious of all dessert apples is Claygate Pearmain, with its green flesh and dull russet-yellow skin, but it is only at its best if removed early in December. Carefully stored it will then keep until March. If removed in October, when it appears to be ripe, it will keep only until Christmas, when it will tend to become soft and lose its crisp nuttiness. Two others which require the same treatment are Cornish Gilliflower and Christmas Pearmain, both apples with their neat habit being ideal for a small garden, and both crop heavily early in their life.

That pleasantly aromatic apple, Duke of Devonshire, is best picked towards the end of October, likewise King's Acre Pippin, whilst Laxton's Superb should be removed in early November, when it would seem to have reached full maturity.

Most early maturing varieties denote when they are ready for picking by parting from the spur if gently twisted, but the keepers hang on tightly and it is left to one's own judgment when to pick. The one exception to the rule seems to be that good keeper Annie Elizabeth, a cooking apple of value which will begin to fall when fully mature and so should be removed the moment it is ripe. The culinary apples, possibly on account of their great weight do tend to fall when mature, but even so should be allowed to hang on the tree for as long as possible. Fruits of that heavily cropping apple Grenadier can be used for cooking from mid-August and will remain without falling until late October, but being such a woefully bad keeper, they should be used from the tree throughout autumn, and the fruit should be allowed to hang until the last to mature are to be used.

Very different is Newton Wonder, one of the most handsome of all apples, which should be allowed to remain unpicked until November, when it will have reached a huge size and be fully mature, and will store in perfect condition until Easter. Lane's Prince Albert, requires similar treatment, but as it does not keep so well, and the longer it can be left on the tree, the longer it will keep in store.

Another apple requiring November picking is the chalk lover, Barnack Beauty, which will then retain its good dessert qualities until March. The new Bowden's Seedling, will also remain shiny skinned and juicy until April if allowed to hang until late November.

Several apples are not at their best until they have been up to two months in store. These are the apples which will keep until spring. The dual-purpose Edward VII is one. It is removed from the tree a deep green colour, which makes quite uninteresting eating. But by February, the skin is turning a rich golden colour. Do not be fooled by this, leave it until early summer when it will be mellow and deliciously sweet and juicy, then eat it just before George Cave or Beauty of Bath is ready on the tree in July. The new Laxton's Rearguard, is similar. It bears a flat fruit which when ripe in early March turns a rich golden colour, flushed with red and russet. It possesses much the same rich flavour as its parent, Ribston Pippin, whose flavour of which Lindley said in 1830 was " not to be surpassed." But allow it to remain on the tree until early December, to ensure that its full flavour and crispness is to be enjoyed in early spring.

FACTORS DENOTING MATURITY

There are factors to be considered apart from the characteristics of various varieties. One is the soil, for in soils that are light and sandy, the fruit will not hang as long as fruit on trees growing in a heavy soil; and trees growing in grass also retain their fruits better than where the soil is being cultivated beneath the trees.

Apples growing in the North are, through fear of adverse weather and frosts, best gathered by mid-November. In the South the late maturing varieties may be left on the trees for another month.

To find some indication as to whether the late varieties are ready for picking, cut a fruit through the centre and if the pips have become hard and almost black the fruit may be said to have reached maturity. Removing from the tree calls for great care, and because apples are not so easily bruised as most other fruits, it does not mean that they can be handled carelessly. A bruised fruit will quickly deteriorate. They should be carefully twisted from the trees and placed in shallow baskets for removal to the store room. The question of wrapping the fruits is always a debatable point. Wrapping in waxed paper certainly helps to preserve the fruit by preventing evaporation, but there is always the danger that some fruits will decay without being noticed. It may be preferable to place the fruits on waxed or grease-proof paper taking care to see that they do not touch each other, then as soon as any decay is noticed the fruit can be used at once.

STORING CONDITIONS

Wherever possible late apples for keeping should be gathered dry. If wet, wait for a windy day to dry off the moisture. Apples should be stored completely dark and as cool as possible for them to keep for any length of time. A too dry room will cause shrivelling, and a place where there is any undue changes of temperature will cause considerable condensation, which will also be a deterrent to their long keeping. Robert Thompson in *The Gardener's Assistant*, published a century ago, suggested covering stored apples with a thin layer of very dry straw to absorb any moisture, which arises from the fruit, and this method has been found to be a great help to long keeping. Edward VII, and Crawley Beauty may be kept until the following summer, by placing them on a layer of dried bracken; and cover them with the bracken too. On wooden shelves in a stone barn, or dry cellar, the fruit keeps almost indefinitely. A dry shed is also suitable but to prevent temperature fluctuations the roof should either be thatched or lined inside.

PRE-HARVEST DROP OF APPLES AND PEARS

Amateur growers may experience a premature dropping of apples and sometimes pears before the fruit is fully mature, and which causes serious damage to the fruit, which will lose its dessert quality and will be useless for storing. Two offenders are apples: Beauty of Bath and Gladstone; and Conference pear. The placing of straw round the trees during June will not only act as a mulch, but will give the fruit some help in breaking their fall. It is however possible to use certain proprietary sprays containing naphthalene-acetic acid, which if sprayed on to the offending trees before the fruit is more than three-parts grown, will restrain their dropping, which is often caused by a warm July and August, but with cold evening conditions; or merely by strong winds. It is generally the early maturing varieties which are the worst offenders, but Cox's Orange Pippin is often troubled by premature fruit dropping, where soil and climate do not completely suit it. The trees should be sprayed on or about July 1st. Beauty of Bath, one of the most popular of the earliest apples to mature, is most often the worst offender, and chiefly on account of its ability to hold its fruit the new George Cave should be planted as a substitute.

PEARS

Pears must be given even greater care in their picking than apples, for not only do they bruise more easily, but will rapidly deteriorate on the tree if allowed to become over-ripe, or if subjected to adverse weather, such as a period of moisture or night frosts.

Generally a pear will be ready to gather when, upon lifting with the palm of the hand and exerting no pressure on the fruit, it readily parts from the spur. This is a better method than following the text book as to the correct ripening periods of the different varieties, for so much depends upon the season and upon the situation of the tree in the garden. Given a particularly warm summer, with periods of prolonged sunshine, such as in 1955, then pears will come to maturity before their usual time, several weeks earlier in the South. But if the fruit does not readily part from the spur when lifted, it means that it is still drawing nourishment from the tree. Yet equally important with the pear is not to allow it to become over-ripe, for not only will its keeping qualities badly deteriorate, but it will have lost much of its fragrance and flavour. The lifting test is more reliable than the colouring of the skin, for this depends so much on the soil, and if waiting for the fruits to attain a certain colouring they may have passed their best. In this respect, the best of all early pears, Laxton's Superb, at its best in a normal sunny season during the first days of August, must be gathered and eaten whilst still pale green, when it will part from the spurs upon lifting. To allow it to become a buttercup yellow colour, which it soon will if left on the tree, will be to lose its flavour. Similarly William's Bon Crétien, the canner's Bartlett Pear, will become dry and of a disagreeable flavour if allowed to hang too long.

HARVESTING AND STORING

All pears, whether to be eaten at once or to be stored must be removed in the palm of the hand, and carefully placed on a wooden tray, which has been lined with cotton wool. Always gather the fruit when quite dry and store in a slightly warmer place than for apples. At all times pears like warmth, a temperature of 45° F. suiting them best, and an airy attic room if dark and a cupboard or drawer is better than a cellar or shed which will be colder. If placed in a cold room, pears will sweat badly and quickly lose quality. So that the fruit will not lose its bloom, the best method is to place them upright on a layer of cotton wool so that they are not quite touching. As with all stored fruit, keep each variety to itself so that the fruit may more easily be seen to reach the correct state for using.

PLUMS

Unlike pears, plums should be allowed to remain on the trees until quite ripe, the longer they will hang, the better will be their flavour. As plums are generally more numerous than pears, the best test as to whether they are ready for gathering is to remove one and taste it. It should be pleasantly soft, juicy and in no way dry, and the stone should readily part from the flesh. It should also possess a rich flavour.

Few gardeners realise that certain varieties will keep just as readily as most pears if removed by their stalk, so as not to damage the ' bloom ' and if placed in a dry, airy room. The older generation of fruit growers would wrap the fruit in pieces of newspaper for they said it kept better this way, but this should not be necessary. Those that will store for several weeks include Coe's Golden Drop; Laxton's Cropper; Laxton's Delicious, bred from Golden Drop and hence its similar qualities; and the grand old variety, Angelina Burdett. It is said that Golden Drop will keep up to nine months in store, but I have never tried it for this length of time.

CHERRIES

Cherries do not keep for long, but like the plum should not be removed until completely ripe, and for ' white ' varieties like Frogmore, it is better to remove one or two and test as for plums, rather than to rely on its skin colouring.

As birds generally make themselves a nuisance with cherries as they ripen, especially with the black varieties, fish netting should be thrown over wall trees, or over young trees which have not made excessive growth, and this should be done as soon as the fruit begins to swell.

Conference pear against a warm wall

New pear, Roosevelt. One of the largest and best keeping of all pears

Pear, Laxton's Foremost. Ideal for the show bench

THREE PEARS FOR SUCCESSION

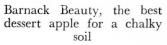

Upton Pyne, a handsome dual-purpose apple

Newton Wonder, a long keeping apple of quality

Barnack Beauty, the best dessert apple for a chalky soil

Blenheim Orange, a valuable dessert apple for a large garden

FOUR VALUABLE APPLES

CHAPTER XI

THE VALUABLE DUAL-PURPOSE APPLES

APPLE LOVERS having only a small garden and who would like to grow as useful, though possibly not the most delicious apples as possible, would be well advised to plant several of the dual-purpose varieties, valuable both for dessert and cooking.

TO USE AUGUST AND SEPTEMBER

Duchess of Oldenburg. I can imagine readers saying, "never heard of it." It is an old Russian apple, extremely hardy and one of the first dual-purpose apples to mature, ready for gathering before the end of August. Old Dr. Hogg wrote of it " an early culinary or dessert apple of first quality," and so it is, ideal for a cold, wind-swept garden. The skin is almost of Jersey cream colour, the flesh being soft and juicy as good cookers should be.

White Transparent. This is another unusual apple which should be more widely planted, for it is dual-purpose and one of the first of all apples to mature, ready in the South at the end of July. It is a strong grower, resistant to disease, and is a heavy cropper, the fruit having a waxy creamy-white skin, the flesh being crisp and sharp, but like Grenadier it should be used from the tree.

TO USE OCTOBER TO MID-NOVEMBER

Charles Ross. A handsome apple of large proportions, the flesh being crisp and refreshing, the skin deep green, heavily flushed scarlet on the sunny side. That it is not a strong grower except where suited, has detracted from its popularity of recent years, but it always does well in a chalky soil. It is also somewhat sensitive to lime-sulphur, and for this reason is not now planted as much commercially as at one time. Nor does it keep too well, becoming rather dry, but taken from the tree mid-October and used before December 1st, there is none better as a dual-purpose apple. Pollinated by most mid-season flowering varieties, especially Grenadier, Ellison's Orange and James Grieve.

Wealthy. This is another dual-purpose American apple of handsome appearance, and which is cropping well in Britain. It is not a new apple but one which is only now becoming popular. It bears a most attractive fruit with a yellow skin, flushed and striped scarlet and russet. The flesh is juicy and refreshing, and is delicious eaten from the tree in November; or it is equally useful for cooking, breaking down beautifully when baked. At its best in November.

TO USE LATE NOVEMBER TO DECEMBER

Forge. Where one's garden is in a frost pocket, or is troubled by cold winds, again where space is strictly limited, plant this old variety which for some reason has almost disappeared. It makes a small, compact tree, comes quickly into bearing and its fruit, with its pale yellow skin and pleasing perfume, will keep from late October until late in January. Hogg wrote, " a great and constant bearer " and yet where is it planted today? If I were a wealthy man I should like to plant a tree in every garden in Britain. It is also highly resistant to scab. What more do we want?

Gascoyne's Scarlet. Also strong growing, but only in a soil heavily laden with chalk. This variety is one of the best of all apples for planting on a thin, chalky soil. The aromatic flesh makes pleasant eating in early November, whilst it is also delicious and soft when cooked. It is a handsome apple with a skin of palest yellow, heavily splashed with scarlet.

Lady Henniker. Raised at Thornham Hall in 1850, this old Suffolk apple now seems to have taken a back seat. It is a fine cooking apple for December, and makes pleasant eating in the New Year as dessert when the fruit has lost some of its acidity. It makes a large spreading tree and crops heavily, the fruit being of large, uneven shape and which keeps well into the New Year.

TO USE JANUARY TO FEBRUARY

Belle de Boskoop. A native of the Low Countries where it is widely planted, this makes a strong growing tree and takes a year or two to come into bearing, but the fruit is so valuable for Christmas use that it could well be more widely planted. That it is not more so is because its bright yellow fruit with its grey russeting is none too attractive. This however, will not put off the amateur, for it makes delicious eating either as dessert or when cooked.

Opalescent. A fine all-round apple of American origin, with its glossy, almost purple-crimson skin and its large size, making it one of the most handsome of all apples. Used for dessert during

December, it is rich and juicy and will remain so until mid-March. It is also, in my opinion, second only to Bramley's Seedling as a cooking apple. Add to this its hardiness and its large and regular cropping habit, it must be an apple with a distinct future. Grown well, the fruits reach huge proportions without in any way becoming coarse, nor do they require thinning. Strange as it may seem, I know only one firm which stocks this variety.

Wagener. Like a number of the dual-purpose apples, this is an American variety and an instance of a variety known for more than 150 years only just becoming popular. It makes a small, compact tree and is an abundant bearer, the fruit being the best keeping of all apples for if taken from storage as late as mid-April, the fruits will not contain a single wrinkle, and will be still firm and juicy. The bright glossy green skin, flushed with scarlet, makes this a most handsome exhibition apple. It needs a pollinator to crop abundantly, preferably Egremont Russet or Lord Lambourne.

Woolbrook Pippin. Introduced in 1920 by Messrs. Stevens of Sidmouth, this is a splendid dual-purpose apple for January to February use. It may be used for cooking until Christmas, but afterwards its juicy and aromatic fruits with their yellow and red russeted skin, are so delicious as to be worthy of best dessert. It is a vigorous but upright grower, and bears especially well in light soils.

TO USE MARCH TO MAY

Crawley Beauty. Found growing in a Sussex garden and introduced by Messrs. Cheal & Co. of Crawley, this is a superb apple and quite indispensable in a garden troubled by late frosts, being in bloom the first days of June. The fruit should be allowed to hang until mid-November, when carefully stored it will keep until April. The skin is deep green, striped and spotted with crimson, the flesh is soft and sweet. The tree is resistant to the usual apple diseases and is of upright habit, making it most suitable for a small garden.

Edward VII. This in my opinion is one of the very best of all apples. Like Crawley Beauty, it blooms late and makes a neat, upright tree. It is the result of a Blenheim Orange x Golden Noble, two splendid apples.

Though usually listed as a cooker, Edward VII, makes delicious eating if kept until Easter, and it will store until the first July apples are ready. It will then be golden skinned and possess a rich, sweet flavour. I once made myself ill with eating so many at the place where I worked, the owner kindly granting permission to eat as many as I wished, as there were so many still to be used! The fruit should be allowed to hang as late as possible.

Lemon Pippin. This strangely shaped variety grows in my garden, making a small, compact tree and bearing a huge crop of medium green fruits quite oblong, almost lemon shaped. But do not be put off by this, even if your neighbours laugh, for its fruits hang until Christmas, and it will store until Easter, its rich, sharp flavour being excellent for cooking and reasonable for dessert.

Upton Pyne. Bearing heavy crops in the worst seasons and making a large apple of exhibition quality is Upton Pyne, raised by Mr. George Pyne of Topham in Devon. The skin is of a pale primrose, striped with pink, and the fruit will keep well until Easter. The flesh is crisp and sweet, whilst it makes possibly the best baked apple of any, Bramley's included. Should also be planted for beautiful blossom.

Apples for exhibition—
 Charles Ross, Crawley Beauty, Opalescent, Upton Pyne, Wagener, Woolbrook Pippin.

Apples for a very small garden—
 Charles Ross, Duchess of Oldenburg, Edward VII, Forge, Lemon Pippin, Wagener.

Apples for a cold, frosty garden—
 Crawley Beauty, Duchess of Oldenburg, Edward VII, Forge.

CHAPTER XII

NEW DESSERT APPLES OF MERIT

IT TAKES at least a quarter of a century to introduce a new apple and at least fifty years for it to impart its merits on the public. Several varieties which were introduced about the year 1930, after trials covering a period of at least twenty-years are only now (1956) becoming popular and being widely planted. Amongst these are Woolbrook Pippin, Howgate Wonder, Laxton's Advance and Exquisite, all apples of proved merit, but to many gardeners, still considered ' new ' varieties.

There has yet to be discovered the perfect dessert apple, one having the flavour and quality of Cox's Orange Pippin, self-fertile and combined with the neat, yet clean habit of Adam's Pearmain, and the reliability of Worcester Pearmain; an apple also having a handsome appearance for exhibition, or for sale, though this is not important with the amateur grower, who may plant chiefly on the quality of the fruit and on the habit of the plant. Claygate Pearmain, for instance, is to my mind the most delicious of all late apples, but does not sell readily on account of its lack of colour, the townsman demanding a highly coloured fruit no matter what the quality. The result has been that during the past twenty-five years many of our most delicious old apples have been entirely neglected by commercial growers, and if we are to enjoy their individual flavour again we shall have to grow our own. There are however, a number of varieties introduced since 1946, over the past decade, which give promise of great things. It is as yet much too soon to make any dogmatic statement about their merits over existing varieties, which have been long established, but each shows considerable promise and could prove both interesting and satisfying to the amateur fruit enthusiast.

Apples in order of maturing—
    Early.
        1. George Cave
        2. Elton Beauty

3. Celia
4. Tydeman's Early Worcester.

Apples in order of maturing—
Mid-season.
1. Worcester Cross
2. Hereford Cross
3. Laxton's Favourite
4. Taunton Cross
5. Michaelmas Red
6. Merton Worcester.

Apples in order of maturing—
Late.
1. Pearl
2. Shaw's Pippin
3. Tydeman's Late Orange
4. Kidd's Orange Red
5. Merton Prolific
6. Acme
7. Laxton's Rearguard.

TO MATURE JULY AND AUGUST

Elton Beauty. This is a most handsome fruit certain to become a favourite for the show bench. It is the result of a cross between James Grieve and Worcester Pearmain, bearing the better qualities of these two prolific cropping apples. Its green skin is flushed and striped bright scarlet, with an attractive green ring round the centre. This is possibly the best flavoured of all early apples, maturing at the end of August, and yet keeping until early December, the only long keeping, early apple. Just right for the late summer shows, and for the late seaside trade, where grown for profit, this is one of the best of all early apples.

George Cave. Is this the long awaited apple to replace Beauty of Bath, with its spreading habit and inability to hold its fruit until fully mature? It matures a week before Beauty of Bath and has quite exceptional fertility, its bloom being resistant to frost. The skin is almost as highly coloured as a ripe Worcester Pearmain, the flesh being white, firm, sweet and juicy, with almost no core. The next generation of gardeners will plant this abundantly.

TO MATURE SEPTEMBER

Celia. Raised by Mr. N. Barritt near Chester, this is proving a valuable late mid-season apple. It was raised from Langley Pippin x Worcester Pearmain, bearing a heavy crop of fruit which stores until early December. This is an apple of beautiful shape, with a glossy green skin, mottled and striped reddish-brown, and is ready for eating about September 1st. The flesh is sweet and crisp, and the fruit is well able to stand up to adverse weather.. Like Mr. Barritt's other introduction, Elton Beauty, this apple is free from mildew, and able to withstand lime-sulphur.

Hereford Cross. Not having grown this variety I can offer few remarks, but in the West Country it is being fairly widely planted. Raised by Mr. Spinks at the Long Ashton Research Station, Bristol, it has Cox's Orange as a parent, the fruit having the same crisp, orange flavour, and is of similar appearance. It is ready for eating at the end of September.

Laxton's Favourite. This new apple, with its high colouring and crisp, sweet flesh is ready for eating towards the end of September. It is a vigorous, but upright grower, and so is suitable for a small garden. It crops regularly, the fruit being of an even size and well shaped.

Tydeman's Early Worcester. Raised at East Malling, the result of Worcester Pearmain x MacIntosh Red, the fruit is mature about ten days before Worcester Pearmain, and has the same glossy crimson skin. The round, medium sized fruit is at its best about the first week of September.

TO MATURE OCTOBER AND NOVEMBER

Kidd's Orange Red. This apple was raised in New Zealand between the two wars but is new to this country. It may be said to be the most highly coloured of all late apples, and is now being extensively planted commercially. It has the same high colour of Worcester Pearmain, and possesses even better keeping qualities than Cox's Orange, moreover, it is one of those valuable varieties which, like Herring's Pippin will crop well with the very minimum of attention. In Essex where it is most successful, it is a great success when grafted onto cookers. It is resistant to scab and makes a neat, upright tree.

Merton Worcester. Raised by Mr. M. B. Crane at the John Innes Institute, the result of a Cox's Orange x Worcester Pearmain, it ripens a week later than Worcester Pearmain, is a better keeper, but has the appearance of Cox's Orange, a yellow skin, flushed with scarlet and russet. The creamy-yellow flesh is crisp and aromatic. Does best in the drier districts, and is being widely planted in East Anglia.

Michaelmas Red. Raised by Mr. H. Tydeman at the East Malling Research Station. The fruit is almost a replica of its parent Worcester Pearmain, having the same shiny crimson skin, and matures about a fortnight later. For a small garden this is a better variety than its parent, for it is of less vigorous habit, and is not a tip bearer, so may be grown as a cordon.

Shaw's Pippin. This is an apple of unknown parentage, found in a garden near to the late G. B. Shaw's home at Ayot St. Lawrence in Hertfordshire. It makes a large apple, of rich colouring, and possesses the flavour of Blenheim Orange. It is ready for gathering mid-October and will keep in condition until the year end. The tree comes more quickly into bearing than Blenheim Orange and is a more suitable variety for a small garden. Although the fruit is large, no thinning is necessary.

Taunton Cross. Also raised by Mr. Spinks at Long Ashton, this is a mid-October maturing apple of most handsome appearance, very similar to Charles Ross though the fruit is flatter, its green skin having a bright crimson flush. The flesh tinged with pink, is particularly sweet. Of dwarf habit, the tree is extremely resistant to scab and crops particularly well in wet districts.

TO MATURE DECEMBER TO JANUARY

Acme. Of all the dessert apples introduced over the past decade, this gives promise of being the most outstanding, and has been named accordingly. Raised by W. Seabrook & Sons Ltd., of Chelmsford, this is the first apple to crop on its own roots, setting its fruit freely, even in the nursery rows the second season after planting.

It is later than Cox's Orange, of which it is the nearest in flavour of any apple, the bright yellow skin being flushed and striped bright crimson on the sunny side. The flesh is yellow, firm and juicy. If gathered mid-November, it will keep through winter. To date it appears to be what the fruit grower has long been looking for, " an easy Cox's Orange."

Merton Prolific. This is a late maturing apple, raised by the John Innes Institute. It is a regular and heavy cropper, the almost olive-green skin having a striking carmine-red flush which becomes brighter with keeping. It has Cox's Orange Pippin and that excellent dwarf late cooker, Northern Greening, as parents. It makes a neat, upright tree. The fruit should be gathered mid-November, and will keep until the end of February.

Pearl. A really good dessert apple for the Christmas period is Rival, one of the parents, with Worcester Pearmain of this new apple. Pearl blooms reasonably late, and is extremely frost resistant, though it is tip bearing and is of vigorous habit. It is a heavy cropper, the conical fruit hanging well and ripening to a deep red colour by late September, the flesh being yellow and with almost a Cox's Orange flavour. It will keep well until the end of January, the flavour improving with storing. One of the best of all apples for the Festive Season.

TO MATURE FEBRUARY TO MAY

Laxton's Rearguard. This is the longest keeping of all dessert apples, only Edward VII, which may be termed a dual-purpose apple, keeping longer. It is an extremely hardy variety, with the same characteristics as its parents Court Pendu Plat and Ribston Pippin, both being hardy and of compact habit. The fruit, which has a slightly russeted appearance, is similar to Cox's Orange, but is of a more flattened shape. It should be allowed to hang on the trees until early December, and is not at its best until March. It will keep until June.

Tydeman's Late Orange. Also raised by Mr. Tydeman at East Malling, this variety may be described as a late keeping Cox's Orange, storing until April, and being of similar flavour with dark crimson-russeted skin. It is pollinated by Grenadier or Charles Ross. Would appear to crop heaviest and remain more free from scab in the drier districts.

CHAPTER XIII

APPLES WITH DISTINCT FLAVOUR

WHERE SPACE is limited it may be preferable to plant apples noted for their rich, individual flavour, for this will enable one to appreciate these fruits as a welcome change from the rather over-planted Worcester Pearmain, etc. In the same way as many gardeners rely on the mass-produced Majestic and King Edward potatoes supplied by the farmer for the bulk of their supplies, but planting such epicurian varieties as Golden Wonder or Dr. McIntosh to provide a welcome treat for special occasions. We have become so accustomed to just one or two varieties, that the present day gardener who may be a fruit lover, remains completely ignorant of those varieties neglected for some reason by the commercial grower. So here is a selection to provide fruit the whole year round, and which will make the storing and growing of fruit extremely interesting.

RIPE IN JULY AND AUGUST

Elton Beauty. See Chapter xii.

George Cave. See Chapter xii.

Irish Peach. A tip bearer and a vigorous grower, and though not suitable for cordons it bears an apple of such delicious flavour, that it should be grown wherever space permits. The fruit, with its crisp, aromatic flesh is ripe the first week of August, and should be eaten during that month. The fruit is of conical shape, is pale green, mottled and flushed with crimson.

Lady Sudeley. Like all early apples this one is extremely highly coloured, being of a rich golden colour, vividly striped with scarlet. This makes the best tree of any for a tub or pot. It bears heavily and remains free from disease. It comes into use early in August, but must be gathered just before fully ripe to obtain its best flavour. Flowering late, it is a valuable early apple for a frosty garden, and especially in the North.

Laxton's Advance. Ripe in mid-August, when it should be eaten, and not left to become dried by the sun, this is a delicious apple

APPLES WITH DISTINCT FLAVOUR

of large, handsome appearance, and possessing the crisp texture of Cox's Orange Pippin. The fruit is coloured a rich crimson, and may be said to be the sweetest of all apples, too sweet for some.

RIPE IN SEPTEMBER AND EARLY OCTOBER

Autumn Pearmain. One of the oldest apples in cultivation, it was once the favourite of Tudor gardeners, and is still an apple of outstanding merit. It makes a large tree, and is extremely prolific. The fruit possesses a rich, nutty flavour, but is frequently too small, hence its neglect.

Devonshire Quarrenden. Another old favourite, grown in the West Country in 1650, and having a brisk, juicy flesh, and a flavour peculiar to itself. It is not too fertile and suffers from scab, but its shining crimson fruits are quite delicious. The tree is hardy away from the West, where it will prove healthier.

Ellison's Orange. Valuable in that it quickly comes into heavy bearing, the large handsome fruit being of Cox appearance, but must be eaten early September, just before it is ripe. If over ripe, the flesh is soft and has a peculiar aniseed flavour. A good Cox pollinator and extremely resistant to scab, mildew and canker.

Laxton's Epicure. Like Fortune, this is another apple of excellent qualities from the Cox's Orange x Wealthy stable, and a winner of the Bunyard Cup for the best seedling apple. It is ripe and should be used during September, when it will be found to possess the juicy sweet flavour of Cox's Orange. Like Fortune it is self-sterile, and needs James Grieve or Worcester Pearmain as a pollinator.

Laxton's Exquisite. The result of Cox's Orange x Celleni Pippin, a handsome dual-purpose apple. It follows Worcester Pearmain, ready at the end of September. It it a vigorous upright grower, the fruit being highly coloured and of rich flavour, but has only a short season before it goes soft and dry.

Laxton's Fortune. Rightly given an Award of Merit by the R.H.S. this is one of the best apples ever introduced, the result of a Cox's Orange x Wealthy. It is early to bloom and is self-sterile, and must be planted with Lord Lambourne, or Laxton's Exquisite. The fruit is at its best during October, when it should be used for it will not keep. It is a strong and regular cropper, the bright yellow fruit being profusely striped with red.

RIPE MID-OCTOBER TO END NOVEMBER

Allington Pippin. Those who enjoy a brisk, acidy apple, as does the author, will find this a welcome change from the rather sweet earlier maturing varieties. It is inclined to biennial bearing, and makes a large spreading tree. Growth is more restricted in poor soils, those of a dry, sandy nature for which it is most suitable. This could be said to be a dual-purpose apple, for like all those possessing a tart flavour, it cooks well.

Herring's Pippin. Introduced by Messrs. Pearsons of Nottingham, at the beginning of the century and is a most reliable apple for a cold, heavy soil, cropping freely if entirely neglected. The deep green fruit flushed crimson on the sunny side, possesses a strong aromatic perfume and spicy flavour. Should be used through November, there being no better apple for this month, yet it still remains neglected.

King of the Pippins. This is an exceedingly hardy variety, which at one time was widely planted commercially. It makes a small, upright tree and bears heavy crops, the orange coloured fruit possessing a distinct almond flavour, the flesh being firm and nutty like a russet. James Grieve, Beauty of Bath or Grenadier, would be good pollinators. The fruit should be used during November.

Mother. An American variety introduced by Messrs. Rivers of Sawbridgeworth. The yellow and crimson fruits with their pinky flesh possess rich, aromatic flavour, and are exceptionally sweet and juicy. The tree makes slow growth though cropping heavily, whilst it is an excellent variety for a Northern garden.

FOR CHRISTMAS DESSERT

To have a selection of homegrown apples on the table for the Festive Season, with their individual flavours is infinitely better than relying on shop fruit, produced in America, or the far corners of the world, and which has lost much of its flavour. These apples provide excellent eating, and may be considered amongst the most highly flavoured of all.

Christmas Pearmain. Like Claygate Pearmain, it makes a neat, upright tree, is extremely hardy, does well in a cold, clay soil, and crops heavily. For an exposed garden it should be included, but though its flesh is crisp and juicy, it cannot compare in flavour to the others of this section.

Claygate Pearmain. Found growing in a hedge at Claygate in Surrey about 150 years ago, this is one of the very finest of all dessert apples, included in my 'twelve best' and Edward Bunyard also honoured it in his best dozen—" fully deserving a place," he wrote. Anyone who has gathered the green and grey russeted fruit, covered with frost on a late December morning, will have tasted the English apple at its very best. The flesh is also green, deliciously sweet and crisp, and very aromatic. It makes a neat, compact tree, ideal for a small garden, for which it would be my first choice. Few know it because it is the vividly coloured imported apples which attracts most attention today.

Cornish Aromatic. Its handsome fruit of orange and red, is marked with russet and possesses a nut-like aromatic flavour. This is a neat, hardy variety, excellent both for cooking and dessert. It is not a heavy cropper, but is so resistant to scab and canker in wet districts that it is still valuable for these areas.

Cox's Orange Pippin. Raised from a pip of Ribston Pippin, which accounts for its quality, by Mr. W. Cox, a brewer at Colnbrook, and introduced in 1850 by Charles Turner, who gave us that excellent cooker Arthur Turner. Cox's Orange not only possesses superb flavour, but is of arresting appearance, which accounts for its popularity. It does however, possess a wide variety of adverse points, which make it one of the most difficult to crop well. The blossom is susceptible to frost, the tree may be termed a weak grower, and is very frequently troubled by scab, mildew and canker in cold soils. It is sensitive both to lime-sulphur and copper sprays. Yet the fruit carries a more subtle blending of fragrance and aromatic flavour than any apple, and nothing has yet been found to take its place, though Acme may do.

Gravenstein. This apple may be included in the previous group for it is ready for eating at the end of October. It will however keep until Christmas, and is of such superb quality that it is a sacrilege to use it before. To enjoy its honey-like flavour to the full, one needs to be able to sit in front of a log fire entirely at ease. It is an old German variety of poor appearance, which conceals its soft, juicy, creamy flesh with its subtle aroma. It makes a huge, spreading tree, requires plenty of room and should be given a warm soil. Not for North Country gardens.

Margil. A very old apple, which Edward Bunyard includes in his best ' selection ' for Christmas. It makes a very small tree and bears heavily in all districts. It is not popular for its small, flattish fruit is not in any way handsome, being yellow and crimson and covered with splashes of russet, but this matters little, it is what is inside that counts, and the crisp yellow flesh is sweet and juicy and strongly perfumed. With a light sherry this fragrance is brought out to the full, but use it by the early New Year. Robert Thompson suggests that it should be grown in wind-swept gardens on account of its dwarf habit.

Ribston Pippin. Though almost past its best by Christmas, being suitable for November and December, this magnificent old apple, found in the garden at Ribston Hall, Knaresborough, about the year 1750, has achieved fame as the parent of Cox's Orange, as well as for its own delicate flavour. It makes a spreading tree and crops regularly, though sometimes lightly, and must have plenty of moisture at its roots. The fruit is most handsome, being of an olive-green, striped and flushed scarlet. Still grown commercially throughout the world, but is now strangely neglected by the amateur. Do not forget it is a triploid.

Wyken Pippin. One of our oldest apples introduced from Holland at the beginning of the 18th Century. It is a compact grower, and comes quickly into bearing in the standard form. The greenish skin is dotted with russet, its yellow flesh being of delicious flavour from December until March, but it needs a moist soil.

TO USE IN THE NEW YEAR

Adam's Pearmain. For the period Christmas to mid-March, there is no better apple for the small garden, and Lindley writing in 1830, said that this apple possessed the finest flavour of any dessert variety. It makes a small tree and is extremely hardy, cropping well even in clay soils. The yellow russeted fruit is particularly handsome, the yellow flesh being rich and juicy.

Cornish Gilliflower. An old variety discovered in a cottage garden near Truro, it is not a heavy cropper, and makes a vigorous straggling tree, and yet the honey-like flavour of its fruit demands that it still be planted. All the old writers sing its praises, yet it is a tip bearer, and unsuited to the small garden.

Court Pendu Plat. One of the oldest apples in the world, widely planted in Tudor gardens, and which is still a valuable variety. Its hard, yellow flesh is pleasantly aromatic, much like Cox's Orange, but unlike that variety, it blooms very late and so misses late frosts. For this reason it was known to Stuart gardeners as the Wise Apple. Though making only a very small tree, it bears heavily, a handsome highly coloured apple.

Easter Orange. This is an excellent apple introduced by Hilliers of Winchester, the fruit, orange flushed with scarlet and russet, being at its best for Easter, the creamy flesh being crisp and sweet. The tree is of quite vigorous growth, but like Claygate Pearmain, is of neat habit.

Heusgen's Golden Reinette. A variety from the Low Countries, the fruit being like a small Blenheim in colour, size and flavour, but comes into bearing much sooner. The tree is hardy, and is a heavy cropper.

King of Tomkin's Country. An American apple of very vigorous habit, making a large spreading tree, and bearing a handsome carmine coloured fruit, at its best during March.

Laxton's Rearguard. See Chapter xii.

APPLES WITH DISTINCT FLAVOUR

May Queen. Raised near Worcester, this is an ideal apple for a small garden, making a very small, compact tree, immune to scab, yet cropping heavily. The fruit with its crisp, nutty flavour is at its best if kept until May.

Sturmer Pippin. It is so late to mature that it should only be planted where the fruit receives the maximum of late autumn sunshine, and it always seems to do best south of Birmingham. It was raised at Harverhill in Suffolk, about a century ago, from a seed of Ribston Pippin, (surely the best and most prolific parent of all apples). Though green the fruit is handsome, being covered with dark brown russet, the flesh being firm and having a gooseberry-like flavour. The fruit will keep until June, and is only at its best early the following summer.

Dessert selection for a small garden or orchard for succession (20 trees for an outlay of less than £10):—

Late July—George Cave or Lady Sudeley
Mid August—Laxton's Advance
Late August—Elton Beauty
Mid September—Michaelmas Red
Early October—James Grieve
Mid October—Laxton's Fortune
Late October—Egremont Russet
Early November—Herring's Pippin
Late November—Ribston Pippin
Early December—Orlean's Reinette
Late December—Acme or Margil
Early January—Claygate Pearmain
Late January—Adam's Pearmain
Mid February—Court Pendu Plat
Early March—Brownlee's Russet
Late March—Laxton's Rearguard
Early April—Easter Orange
Late April—Edward VII
May—May Queen
June—Sturmer Pippin.

All are hardy, easy to grow and suitable for cordon culture.

CHAPTER XIV

DESSERT APPLES FOR ORCHARD AND LARGE GARDEN PLANTING

For large or small orchard planting, and where there is room available in the garden, those who wish to grow for an abundance of fruit and possibly profit, will find the heavily bearing varieties, which make large, vigorous trees, many being tip bearing and so are unsuited to growing in the artificial form, more suitable than those having a dwarf habit. So much depends upon the size of the garden. For instance, where given room to spread, there is no more delicious early winter apple than Gravenstein, a " must " for every large garden, whilst the raspberry-like flavour of Worcester Pearmain, would always be placed in the first dozen of all dessert apples by the author, yet here again, it makes a large tree and is tip bearing. And whilst it is the most popular of all apples for pollinating Cox's Orange, grown commercially, a more compact tree would be better for the small garden. Again, many of those with vigorous habit, e.g. Blenheim Orange, Bramley's Seedling, Gravenstein, Allington Pippin, take several years longer to come into heavy bearing, and though they will more than make up for this with extraordinary heavy crops in later years, the amateur will generally wish to enjoy his fruit as quickly as possible. Where there is room in the garden, or where a small orchard may be planted, then these vigorous growers are essential, and should be planted on a suitable rootstock, not of dwarfing habit.

TO RIPEN JULY TO AUGUST

Beauty of Bath. Introduced about a century ago by Messrs. Coolings of Bath. It makes a vigorous, spreading tree and has been widely planted until recently on account of there being no similar variety, bearing highly coloured fruit so early. In my Somerset orchard, this variety is ready by late July when it should be used, otherwise the fruit will become dry and flavourless. It would appear that George Cave and Laxton's Advance would re-

Three years old Dwarf Pyramid apples

Cordon apples and 10-tier horizontal pears forming a delightful fruit walk'

Laxton's Superb, a fine pear for tub or pot culture

Two year cordon apples, Laxton's Epicure

FRUIT IN TRAINED FORM

place it for commerical planting. It blooms early and may prove useless in a frosty garden, whilst it is tip bearing and self-sterile. Use Laxton's Advance as a pollinator.

Gladstone. This is the first of all apples to mature, being ready for eating mid-July in a sheltered garden. It makes a large, spreading tree and is a tip bearer, whilst it bears very large fruit, highly coloured, and which must be used just before it is mature or it will become soft and flavourless. The variety was found growing in Worcestershire and introduced about the same time as Beauty of Bath.

Irish Peach. See Chapter xiii.

TO RIPEN IN SEPTEMBER

James Grieve. Except in the warm, moist districts of the West, where it cankers badly, this is one of the most reliable of all apples. It should not be over-fed with nitrogenous manures, or it will make excessive growth and crop less freely. It is early to bloom, and should not be planted where late frosts persist. Gladstone or Laxton's Epicure are suitable pollinators, whilst James Grieve is an excellent pollinator for Cox's Orange. Though it is of vigorous habit, it is of upright growth, and as it is a spur bearer proves suitable for all but the smallest gardens. It crops well and regularly, the fruit being of a rich flavour from the tree early in September, or stored for 3–4 weeks. One of the few apples introduced from Scotland, and valuable in every way.

Miller's Seedling. Raised by James Miller of Newbury, this is an excellent early September apple, at its best just before Worcester Pearmain. It makes a large tree, yet comes quickly into bearing and crops so heavily that it tends to biennial bearing, requiring a season of rest after one of plenty. It is a handsome apple with pale yellow skin, striped scarlet with the fruit on the small side and with no outstanding flavour. It will prove reliable in all soils, and I have also seen it crop well in a chalky soil.

Worcester Pearmain. Unlike the equally highly coloured Gladstone, this apple is generally picked all too soon or as soon as it colours. The quality will be greatly improved if allowed to hang for several weeks, when it will be as crisp and juicy as the best of dessert apples. It may be said to be the best apple of its period, and is widely used as a Cox's pollinator, but it is a tip bearer and makes a large tree, and so should be omitted from the smallest gardens, even though it may be said to be the best all-round apple ever introduced, good in all soils, completely hardly, a regular cropper and free from disease.

TO RIPEN IN OCTOBER

Lord Lambourne. Raised by Messrs. Laxton Bros. and awarded the R.H.S. Cup for the best seedling apple of 1921, and since then has been widely planted commercially. It makes a good sized tree and does especially well in standard form. It is self-fertile and is a heavy cropper, but is best away from damp districts, like James Grieve, from which it was evolved, and being of similar colour. It is ripe mid-October and remains sweet, crisp and juicy until late in November. A grand apple for a large garden and should be in every collection.

TO RIPEN IN NOVEMBER

Rival. This is one of the best of all apples for November, being of outstanding flavour, no apple being more juicy. It makes a large, spreading tree and crops well in the North, the handsome fruit having an olive-green skin, flushed bright scarlet on the sunny side. May be eaten from the tree mid-October to mid-November, or may be stored until Christmas.

TO RIPEN IN DECEMBER

Blenheim Orange. One of the great apples of England in every sense of the word. It makes a huge tree, bears a tremendous crop and one of the largest of dessert fruits. It is at its best from mid-November until early January, useful both for dessert and for cooking, like eating sweet Brazil nuts. It was discovered at Woodstock, near Blenheim, Oxfordshire, about 150 years ago, and as long ago as 1822 was awarded the Banksian Silver Medal by the then London Horticultural Society. It has however, two disadvantages. One is that it is a triploid and though being pollinated by James Grieve, Ellison's Orange, etc., is not able to pollinate them in return. Another, is that it takes ten years to come into heavy bearing, and is therefore little planted in private gardens. And like its culinary counterpart, Bramley's Seedling, its blossom is most susceptible to frost.

Laxton's Superb. Like James Grieve and Worcester Pearmain, this is a grand all-round apple, making delicious Christmas and New Year eating. Like its parents, Wyken Pippin, it is extremely hardy, much better than Cox's Orange, for a Northern garden, and much easier anywhere, and yet it has the flavour and high quality of Cox's Orange, its other parent. Though widely planted as an orchard tree, it crops abundantly in the cordon form, and makes a compact bush tree, ideal for any garden. It is a tremendous cropper, the fruit being slightly larger than Cox's Orange, of similar colouring but with pure white, nutty flesh. Carefully stored

DESSERT APPLES FOR ORCHARDS AND LARGE GARDENS

it will keep until March. Pollinated by Rival, Worcester Pearmain and others. Like all huge croppers, often requires a rest season, and is inclined to biennial bearing.

Sunset. With its sweet, yellow flesh and orange russeted skin, this is an excellent substitute where Cox's Orange proves a poor grower. Raised in Kent in 1920, Sunset has all the good points of Cox's Orange, and none of its bad ones, and whereas Cox's likes a sandy loam, Sunset likes a heavy loam. The tree is strong growing without being too vigorous, the blossom is very fertile, whilst it bears heavily from an early age. It also blooms late and so misses late frosts. Making top quality dessert for November and December, this is an apple with a real future.

TO RIPEN JANUARY TO FEBRUARY

Barnack Beauty. For a chalk soil this is the best of all dessert apples, but only on such soil does it crop abundantly. It was originally a seedling found in the village of Barnack, near Stamford in 1900. It is a tip bearer, and makes a huge, spreading tree. The fruit is very handsome, being of beautiful shape and of a deep golden colour, heavily flushed with crimson. The flesh is yellow, juicy and extremely sweet, at its best during January and February.

TO RIPEN MARCH TO APRIL

Winston. This is a fine late keeping apple, raised in Surrey, in which county and in Sussex it crops extremely well. It is also valuable for a Northern garden, for it blooms quite late, missing all but the latest frosts. The fruit is one of the richest coloured of all apples, being bright orange profusely streaked with scarlet, the flesh being sweet and having a strong aromatic flavour. It makes a compact tree, is immune to disease and bears consistently well. It is in fact one of the best apples now grown commercially, and should be more widely planted in private gardens.

A number of apples described in previous chapters are also suitable for orchard and large garden planting:

Lady Sudeley	August
Laxton's Epicure	September
Laxton's Fortune	October
Allington Pippin	October
Charles Ross	October—November
Gascoyne's Scarlet	November
Egremont Russet	November
Gravenstein	November—December
Cox's Orange Pippin	November—January

Claygate Pearmain	December—March
Upton Pyne	December—April
Lady Henniker	January—February
Crawley Beauty	January—April
Edward VII	January—June

These apples are tip bearers and are not generally obtainable in the artificial forms (with the exception of Lady Sudeley):

Barnack Beauty	Lady Sudeley (partly)
Beauty of Bath	Pearl
Cornish Gilliflower	St Edmund's Russet
Gladstone	Worcester Pearmain
Irish Peach	

CHAPTER XV

THE RUSSETS

AFTER A long period of unpopularity, russets are back to favour, not only because of their excellent keeping qualities and their own particular flavour, but because of their ability to stand up to all weathers and most soils. Tending to produce fruit on the small side, they must be grown well to be a success. But they do not bruise easily, which makes them the ideal apple for the home fruit grower, who generally has to make use of anything but perfect storage conditions. The russet is at its best over Christmas when it is most appreciated, for it is then richly flavoured, juicy and nutty. Around the log fires during the Festive Season, russets are always most in demand with our household, who seem to live and thrive on apples of every description.

FOR OCTOBER TO NOVEMBER

Egremont Russet. For the small garden or orchard, this is an ideal variety, for it makes a small, upright tree, and crops heavily in all seasons. It also makes a fine cordon. It is at its best a little earlier than most russets, in November, and does not need storing to bring out its flavour and abundant juice. With its round, even shaped fruit, this is one of the most handsome of all the russets, and one of the most delicious for late October dessert.

Pineapple Russet. With its rich pine flavour, this old variety makes delicious eating for early November. It makes a small tree, is very fertile and the fruit, though on the small side, is of similar shape and colouring to Brownlee's Russet. It does well as a cordon.

St. Edmund's Russet. Raised in Suffolk where it seems to make more growth than in the West, it is a tip bearer, and a strong grower. It is early flowering and pollinated by Beauty of Bath. To many it is the best flavoured of all October apples, almost equal to a Cox's, being juicy and sweet. The skin is bright orange, shaded with russet. Only in size and appearance have russets any bad marks against them, in all other respects they are the most hardy and easily managed of all top fruits, and no apples make for better dessert.

FOR DECEMBER TO JANUARY

Cockle's Pippin. Like so many russets, at its best over Christmas, this is an extremely hardy variety. The fruit is sweet and aromatic, the skin being almost an orange colour, shaded with dull russet-brown. Its fruit reaches a good size only in a heavy loam, and should be thinned. Raised in Surrey about 1800, and at one time extensively grown in Surrey and Sussex.

D'Arcy Spice. At its best in Eastern England, for it does best in a sandy soil and in dry areas. It tends to biennial cropping and its fruit is nobbly, not nearly so attractive as the others mentioned, but it is the nearest apple to nut-like eating that I know, is sweet and aromatic, a grand amateur's apple, at its best during December.

Franklin's Golden Pippin. Listed in Hogg's *Fruit Manual* and now grown for sale by only one nurseryman. The skin is a bright yellow, dotted all over with russet. The fruit is small, the flavour brisk and aromatic, at its best from late October until early February. Hogg called it "a first rate dessert apple" and for those who appreciate a good russet it is so.

Orlean's Reinette. This is possibly the most richly flavoured and sweetest apple in cultivation, grown at least 200 years ago, and originating from the Low Countries, where it has for long been popular. The fruit is flat and of a beautiful golden colour, shaded crimson with a large open eye, rather like a small Blenheim Orange. It is at its best over Christmas, " as a background for an old port it stands unapproachable," says Edward Bunyard, in his *Anatomy of Dessert*. It makes a compact tree and is extremely hardy, and like all russets is rarely troubled by disease.

Rosemary Russet. "Very juicy, sugary and highly aromatic," is the description given to this fine late apple by Dr. Hogg. Its skin is golden, tinged with green and red, and covered with brown russet. It was widely planted at the beginning of the 19th Century, when it was regarded as the best of all New Year apples. It is a hardy variety, and makes a small tree.

Sam Young. For cold, clay soils, this is a most reliable apple, at its best from November until mid-February. Its bright yellow skin is russeted with grey, and spotted with brown, the flesh being green and especially rich and juicy. Introduced from Ireland about 200 years ago, it is also known as Irish Russet.

FOR FEBRUARY TO APRIL

Brownlee's Russet. Another fine apple this, which retains its olive green colour when mature. Its flavour is brisk and aromatic, and it keeps in condition right until early spring without its skin shrivelling. The tree is extremely hardy, of compact upright habit, is very fertile and almost completely devoid of disease. Where the soil is cold and none too well drained this is an indispensable keeping apple. I have seen trees standing in water for weeks without any ultimate loss of crop. Its blossom is amongst the most beautiful, being of a rich cerise-pink colour.

Golden Russet. Of all the apples in my garden which yield well year after year, none is more reliable than Golden Russet, of which H. V. Taylor in his famous *Apples of England* dismissed in three lines, but of which Dr. Hogg wrote in his *Fruit Manual* published 100 years ago as being " a first rate apple in use from December to March, crisp, rich and aromatic "—a perfect description. It has a handsome skin completely covered with golden russet. It is an exceedingly hardy variety, and if it has a fault it is that it bears too heavily and should be thinned.

Powell's Russet. Grown in the West Country for over two hundred years, this variety makes a small, upright tree, and comes quickly into bearing. For a moist district there is no better russet, the small golden russeted fruit keeping until June.

Varieties for a small garden—
    Rosemary Russet, Powell's Russet, Brownlee's Russet, Pineapple Russet, Egremont Russet, Orlean's Reinette.

Varities for a cold garden and heavy soil—
    Sam Young, Rosemary Russet, Orlean's Reinette, Cockle's Pippin.

CHAPTER XVI

APPLES FOR THE KITCHEN

WITH FLOUR in the bin, potatoes in the clamp, and apples to use from the trees and from the store, no one need go hungry. In fact, in our large household apples are used for cooking at least once every day of the year. Not only are they enjoyed by the children in so many ways, especially as used for filling pancakes and served with cream and honey, but they play an important part in their health, and also mean a great saving in the family budget. They are always stored with the maximum of care, even though in plenty, for we realise just how valuable they will be, especially during the mid-winter months when cut off by the snow. The apple store is the most useful room about the home, and always treated with the utmost respect, for nothing is more useful than the apple crop, and remember that every large Bramley's Seedling, weighing on an average 1 lb. or more will be worth a shilling if purchased at the greengrocer's.

Every garden should have its cooking apples, and though where space is strictly limited it may be advisable to plant the dual-purpose varieties, it should be said that though of extreme value, they do not make the most choice dessert eating, nor are they the best cookers. They are a compromise, and where there is room to plant enough apples to supply the household, then a selection of the best cookers should also be planted.

Again, for a cold soil and a Northerly aspect, all the cookers with the possible exception of Bramley's Seedling, will prove hardier and heavier bearers than will many of the dessert varieties. The cookers have been much neglected of recent years, yet they are indispensable to the gardener with a family.

FOR JULY AND AUGUST

Early Victoria. Also known as Emneth Early, for it was discovered at Emneth and introduced by Messrs. Cross of Wisbech, at the turn of the century. The bright green irregular shaped fruit is borne in such profusion that in some seasons it may require thinning. The tree is compact, and comes early into bearing.

Sowman's Seedling. For mid-August, this is a useful variety, raised in Lancashire where it crops well, the fruit being very large and round, and of excellent quality when cooked.

FOR SEPTEMBER

Arthur Turner. A vigorous but upright grower, bearing handsome blossom and a large handsome fruit, the polished green skin having an orange flush. The fruit should be used from the tree, but it hangs well and may be removed when required, from late August until November. For this reason it is extremely useful for the small garden.

Grenadier. If only its massive, exhibition quality fruit would keep even a few weeks, this would be one of the finest of all apples, for not only does it make a small, well shaped tree able to withstand any amount of moisture at the roots, but it is a tremendous cropper in all parts, and is used for pollinating a wide range of apples, including Bramley's Seedling, Laxton's Superb, Cox's Orange, Ellison's Orange, and others too numerous to mention. A fine apple in the West Country being highly resistant to scab and canker, its fruit being amongst the best of all for baking, and yet it will not keep. It must be used from the trees. But for this no other apple would be grown for cooking.

Pott's Seedling. Raised in Lancashire, home of so many cookers, this is a very hardy variety of excellent culinary value, but its great value lies in its hardiness and ability to crop well in poor soils.

FOR OCTOBER

Lord Derby. Another excellent Lancashire culinary apple, raised at Stockport, and like all the early cookers, should be used from the trees. It makes a tree of vigorous growth though of upright habit, and bears heavily in all seasons and in all soils. Like Grenadier, it always does well in a wet, clay soil. It bears an irregular green apple, whose flesh cooks to an attractive deep claret colour, and is especially delicious sweetened with brown sugar.

Rev. W. Wilks. Yet another famous offspring of Ribston Pippin, introduced about 1900, and what a handsome apple it is, making large size and being a universal winner on the show bench. The skin is primrose-yellow, thinly striped with scarlet, the flesh is creamy and juicy and quite sweet. Fruits will frequently weigh more than 2 lb. each. The tree is of neat habit and is a tremendous cropper, but the fruit must be used by the end of November.

FOR NOVEMBER

Monarch. Though often used in late November as harvested, this apple will keep, if stored carefully, until April. It was introduced by Messrs. Seabrooks of Boreham, Essex, in 1913, and is a useful Cox's pollinator. The blooms are extremely resistant to frost, but the tree often suffers from brittle wood requiring its lower branches to be supported. A most handsome apple with its olive green and pink flushed skin, it is delicious when cooked.

Peasgood's Nonsuch. Yet another famous apple from the Stamford district of Lincolnshire, as many delicious apples having come from Stamford as there are lovely churches in the town. It makes a dwarf, small garden tree, yet bears a reasonable, though not enormous crop of handsome, golden fruit, with a bright crimson cheek. The tree is very hardy, though it crops well only in a deep, well drained loam.

FOR DECEMBER

Forge. See Chapter xi.

Golden Noble. This fine old Norfolk apple may be used between mid-October and Christmas, its soft yellow flesh being soft and juicy, delicious when baked. It makes a compact tree, ideal for small gardens, its fruit being amongst the most handsome of all, with its orange-yellow skin slightly speckled with grey and brown.

Howgate Wonder. This is a new cooking apple, now being widely planted commercially. It was raised in the Isle of Wight, but is thoroughly hardy, the fruit being ready mid-October and keeping until early February. The large apples are of a handsome green colour, striped scarlet, an excellent exhibition variety.

FOR JANUARY

Bramley's Seedling. One of the richest apples in vitamin C. content, making a huge orchard tree and bearing heavily, where the blossom is not worried by frosts, and where the soil is a well-drained loam. It also takes several years to come into heavy bearing, and it is a triploid, requiring a pollinator and yet being of little use itself for pollination. This is not a variety for an amateur's garden, for it is also a biennial cropper. Against all this, it bears the finest of all cooking apples, which should be used from Christmas until Easter.

Lane's Prince Albert. Raised at Berkhamsted exactly a century ago, this makes a dwarf, yet spreading tree, with drooping branches, and is most sensitive to lime-sulphur. It requires a rich, deep loam when it will bear profusely a handsome apple with white, juicy flesh.

FOR FEBRUARY TO APRIL

Newton Wonder. With its highly coloured fruit, at its best when grown in grass, its pale green skin flushed and striped with scarlet, this is one of the very best cookers for storing. It makes a large spreading tree, and is definitely biennial, but it blooms late and misses late frosts. Plant with Lady Sudeley, Charles Ross or Early Victoria. Raised at King's Newton near Derby, in 1887, the particularly handsome fruit keeps well until Easter.

Northern Greening. A grand small garden cooker, keeping well until mid-April. It makes only a dwarf, upright tree, yet crops abundantly, the fruit being of a rich, glossy green colour. Ideal for a cold garden, its only fault is that the fruit is small, but against this, it never shrivels when stored.

FOR APRIL TO JUNE

Annie Elizabeth. Raised by Messrs. Harrisons of Leicester, this apple may be considered one of the very best for a small garden. It makes a healthy, compact tree, and comes quickly into bearing, the fruit, if harvested at the end of November keeping until Early Victoria is available in early July. Valuable in the North in that it blooms late, it bears a handsome ribbed fruit very popular on the show bench.

Edward VII. See Chapter xi.

For an all year round supply of cooking apples for a small garden—

Early Victoria	July—August
Grenadier	September
Rev. W. Wilks	October
Arthur Turner	November
Golden Noble	December
Northern Greening	January—March
Annie Elizabeth	March—July

Varieties for a cold garden and heavy soil—

Sowman's Seedling	August
Grenadier	September
Lord Derby	October
Arthur Turner	November
Newton Wonder	December—February
Northern Greening	January—March
Monarch	Until April
Edward VII	Until June

CHAPTER XVII

PEARS THROUGHOUT THE YEAR

Though John Scott in his *Orchardist* published in 1860, tells us that he then grew over eighteen hundred varieties of the pear at his fruit farm in Somerset, only a few are now planted commercially in England. It is felt however, that with the commercial grower of pears being compelled to plant only those producing the most outstanding fruit, for size and flavour to enable him to compete with importations from the Mediterranean countries, where the climate is much more suitable for the pear, we have come to regard these one or two varieties as the only pears worth growing. This is far from the case, there being numerous varieties much better suited to the amateur's garden than the temperamental Doyenne du Comice and Seckle, and which crop more readily, and prove hardier and more resistant to disease. It is of course necessary to give detailed consideration to pollination and in the past this has not been done, with disappointing results. As we have seen, those two magnificent pears, Laxton's Superb and William's Bon Cretien, so often planted together are incompatible. But where one's garden receives its fair share of sunshine, and especially where situated in the South, there is no fruit to equal the pear, grown where it can be thoroughly ripened, but always it must be given a position where it may receive some sunshine, or else it is better not planted at all.

TO RIPEN JULY AND AUGUST

Jargonelle. It is of straggling habit, and is a triploid variety, hence the need for pollinators, but it is extremely hardy, is highly resistant to scab, and bears a heavy crop of long tapering fruit, with its own musky flavour. Excellent pollinators are Durondeau and William's Bon Cretien, the former ripening in October, the latter in September. Plant with them Laxton's Superb, and the huge Roosevelt for Christmas, and you will have a succession of fruit.

Laxton's Early Market. Though a new variety, this has already established itself as the best pear for late July, and like all early apples and pears, should be eaten from the tree. The medium sized fruit, with its yellow skin, flushed with scarlet, possesses a delicious perfume. It blooms early and in an exposed garden may be troubled by late frosts.

Laxton's Superb. This is one of the best pears ever introduced, and makes superb eating if gathered mid-to-late-August, and allowed to stand 48 hours in a warm room before eaten. But it must be harvested just as the green skin takes on a yellow tinge; if left later, the quality will have deteriorated badly. The great value of this pear is that though ripening early, it blooms late and so is very suitable for a Northern garden. Good in the bush form in a small garden, for it is of upright habit.

TO RIPEN EARLY-MID SEPTEMBER

Beurré d'Amanlis. Valuable for its hardiness, for it will bear well even in the Pennines. It makes a tree of vigorous, straggling habit, requiring plenty of room, and bears a medium sized russeted fruit of rich perfume. Raised at Amanlis in France about 1795. It is a triploid, and should be planted with Beurré Superfin and Beurré Bedford.

Dr. Jules Guyot. A valuable pear for less favourable gardens, for it blooms late yet crops heavily and acts as a good pollinator for most varieties. It bears a large fruit with yellow skin dotted with black, and should be eaten from the tree. At its best in the dry climate of South East England.

Gorham. A new American pear, very fertile and which makes a neat, upright tree. It is similar in size and colour to William's Bon Cretien, the fruit retaining its pure white colour when bottled. Highly resistant to scab.

Triomphe de Vienne. This is one of those hardy, reliable pears, now quite neglected. The fruit is not large, but is of brilliant colouring and rich flavour. The tree is of dwarf habit and bears a large crop season after season. An excellent variety where those of more temperamental habit prove difficult.

William's Bon Cretien. Possibly the best all-round pear ever introduced. It makes a strong growing, yet compact tree, and bears a heavy, but not regular crop. Introduced as long ago as 1770, by a nurseryman named Williams of Turnham Green, and half a century later into America by Enoch Bartlett, hence its canning name, the pure white flesh being of melting, buttery texture.

TO RIPEN LATE SEPTEMBER TO MID-OCTOBER

Beurré Bedford. Making a neat, upright tree, ideal as a pyramid and bearing heavy crops of glossy primrose-yellow fruits, it is a self-fertile variety, and ideal for a small garden. Raised by Messrs. Laxton Bros.

Beurré Hardy. Making a vigorous, upright tree, especially suited for orchard planting, this is a hardy variety, and a most reliable cropper. The fruit is unique in that the flesh is rose tinted and also carries a delicate rose perfume.

Beurré Superfin. This should be grown where Doyenne du Comice proves difficult, for its golden fruit possesses almost the same quality and flavour. It is quite hardy, but blooms early and may suffer from frost.

Bristol Cross. Raised by Mr. Spinks at Long Ashton, and has quickly become a favourite in the moist climate of the West Country. The fruit, with its bright yellow skin covered with russet, is juicy and sweet, whilst it crops heavily just before Conference.

Laxton's Foremost. A magnificent pear for late September, and a fine exhibition variety, with its clear primrose-yellow skin. It is an upright grower and crops freely, the fruit having buttery flesh, in no way gritty. Ideal for the small garden, and crops well on a West wall.

TO RIPEN LATE OCTOBER—END NOVEMBER

Conference. Most valuable as a pollinator (except with Beurré d'Amanlis), for all mid-season flowering pears, and bearing one of the most delicious of all fruits, its dark green skin being extremely russeted. It is reasonably hardy and no pear crops more regularly.

Doyenné du Comice. With its deliciously melting, cinnamon flavoured flesh, this is the oustanding variety of all pears, but so difficult to crop. It makes a spreading tree, must be given a warm position and a soil well enriched with humus. It likes its feet in moisture, its head in sunshine. Pollinated by Bristol Cross, Beurré Bedford and Laxton's Superb.

Durondeau. Raised in 1811 by the Belgian bearing the same name, it makes a compact tree, and is extremely hardy, but bears better on the Western side of Britain, for it likes plenty of moisture. If gathered at the end of September, the handsome, golden fruit with its crimson cheek, will keep until the end of November.

Emile d'Heyst. Extremely hardy and suitable for a Northern garden, but should be grown in bush form, on account of its spreading, weeping form. The Lane's Prince Albert of the pear world. The richly flavoured fruit with its strong rose perfume should be eaten from the tree late in October, as it does not store well.

Laxton's Record. One of the best November pears, which should be grown where some of the others prove difficult. The medium sized fruit has a yellow skin, flushed with crimson and russet. The flesh is juicy and melting, with a powerful aromatic perfume.

Laxton's Satisfaction. A high quality pear with same parents as Superb. It is fertile, crops heavily, and bears a very large fruit of rich flavour. It makes a tree of vigorous, but extremely upright habit.

Louise Bonne. This pear is a strong grower, making a large well-formed tree, and is a heavy cropper in the warmer districts, especially in the South-West. The green fruit, with its crimson flesh being of outstanding flavour. Grown at least 300 years ago.

Pitmaston Duchess. The large golden russeted fruit, is of exceptional flavour and is a favourite for exhibition, but it is a shy bearer, and a so vigorous grower for a small garden, and takes longer to core into bearing than most pears. It is also a triploid variety.

RIPE IN DECEMBER TO JANUARY

Glou Morceau. To ripen correctly, it must be given a warm, sunny position, and although blooming very late, and being quite hardy, it does better in the South, where it acts as a good companion and pollinator for Comice. The fruit is extremely juicy, and free from grit, and should be eaten early December.

Packham's Triumph. A New Zealand introduction, with a great future and already it has proved itself on a commercial scale in Britain, the fruit keeping in perfect condition from mid-October until late in December. It is a vigorous grower, and a free bearer, the fruit being similar in both appearance and flavour to Comice, and without that pear's difficulties in culture.

Roosevelt. This is the largest of all pears, and the most handsome fruit in cultivation, the smooth golden-yellow skin being tinted with salmon pink. It is a vigorous, but erect grower and free bearer, the fruit being at its best in December.

Santa Claus. For eating in the New Year, this is one of the best pears. The fruits are almost as large as Roosevelt, of delicious flavour, and with an attractive, dull crimson russeted skin. The tree is of vigorous, but upright habit, and is a free bearer, proving with its resistance to scab, extremely useful in districts of moist climate.

Winter Nelis. Exactly the same remarks may be used for this variety as for Glou Morceau, as to its culture. It makes only a small fruit, but its flavour is outstanding, rich and melting, having the perfume of the rose, and will store until February.

RIPE FEBRUARY TO APRIL

Bergamotte d'Esperen. The best variety evere raised by Major Esperen, and should be given the warmth of a wall to ripen and mature its fruit, which with its pale yellow skin remains rich and sweet until March.

Beurré Easter. The richly musk-scented fruit will store in perfect condition until Easter. It is hardy, and a heavy cropper, but requires careful culture throughout.

Catillac. A late bloomer, vigorous grower, and extremely hardy, cropping heavily and requiring plenty of room. The huge crimson-brown fruit should not be harvested until November, and carefully stored will keep until May. It is used chiefly for stewing, but will make pleasant eating during spring. It is a triploid, and should be planted with Beurré Hardy and Dr. Jules Guyot, for pollination. At least 300 years old.

Josephine de Malines. It bears a heavy crop of small, though deliciously flavoured fruits, at their best during February. It is a hardy variety, but prefers the warmth of a South or West wall if planted in the North. It is a regular bearer, but is of rather weeping habit, not always easy to manage. Raised in Belgium by Major Esperen, John Scott describes it as being " juicy and sugary, with the perfume of the hyacinth . . . one of our most delicious pears."

Olivier de Serres. Raised at Rouen a century ago, this little known pear possesses fine keeping qualities. The fruit is deep olive coloured, covered with patches of fawn, the flesh being sugary and juicy with " a savoury perfume," to quote Scott. It makes a small tree, but I have not grown it away from the West Country, so do not know of its hardy qualities.

Hardy Varieties in order of ripening—

Laxton's Superb	Beurré Hardy
Jargonelle	Durondeau
Dr. Jules Guyot	Emile d'Heyst
William's Bon Cretien	Catillac
Beurré d'Amanlis	Josephine de Malines

Pears of spreading or weeping habit—

Beurré d'Amanlis	Emile d'Heyst
Catillac	Josephine de Malines

Varieties of dwarf habit—
    Beurré Bedford
    Beurré Superfin
    Dr. Jules Guyot
    Laxton's Foremost
    Laxton's Superb
    Olivier de Serres
    William's Bon Cretien

Varieties requiring warm conditions—
    Bergamotte d'Esperen
    Doyenne du Comice
    Glou Morceau
    Marie Louise
    Olivier de Serres
    Thompson's
    Winter Nelis

CHAPTER XVIII

FRUIT GROWING IN POTS AND TUBS

Method of growing—Moisture—Requirements—Suitable varieties—Planting—Culture.

WHERE SPACE is at a minimum, apples, pears and plums may be enjoyed by growing in large pots or small tubs. Previously I have described how it is possible to grow choice fruit round the walls of a tiny courtyard, and it is also possible to enjoy fresh fruit on a terrace or veranda, provided the trees receive some shelter from strong winds and are able to receive a liberal amount of sunshine. Apples and pears are more easily managed under these conditions of restricted planting than plums, and where town culture is required, then apples would prove more reliable and dessert apples rather than cookers on account of their size.

METHOD OF GROWING

It is usual to grow in pots in the single cordon system, in this way the maximum number of different varieties may be grown. This will considerably help with pollination where bees and insects are generally few, besides providing the maximum weight of fruit from the minimum of room. The cordons should be supported by stout canes, which when growing against a wall, will be held in position by strong wires looped round each cane and fastened to the wall at 7–8 ft. intervals by a strong nail.

Where a wall, especially a sunny wall can be provided, this will prove ideal for all fruits, for not only will the trees be protected from strong and cold winds, but the fruit will ripen and colour better than it would where growing in the open ground. If a wall cannot be provided and the plants are to grow unprotected, it will be better to grow several dwarf pyramids in tubs, for they will be better able to withstand strong winds.

FRUIT GROWING IN POTS AND TUBS 115

Both the horizontal form for pears and the fan-shaped tree for plums may be used against a wall, whilst pears and apples may be grown in the single or double cordon form. Besides the necessity for the need for particular care to be taken in the selection of suitable pollinators, the most suitable trees will be those which form close spurs, rather than those which bear fruit on the tips of the wood, and which are of less compact habit.

MOISTURE REQUIREMENTS

Another matter of the utmost importance is to provide sufficient moisture, lack of which is the one chief cause of failure with fruit trees in pots or tubs. Lack of moisture will prevent the fruit from reaching its normal size, without which it will lack flavour and may not store well, whilst the fruit may also fall long before it is mature.

Any plant growing in a pot or tub will dry out at the roots during the period June to September, far more quickly than will a tree growing in the open ground, which may be provided with a mulch to retain moisture in the soil, as well as being able to search more freely for its food and moisture. It must also be remembered that a tree in a pot or tub will have its roots subjected to the almost unprotected rays of the hot summer sun. It is therefore imperative that the maximum of protection is provided for the plants and during May straw, strawy manure or sacking should be packed around the pots and kept always damp. This will protect the rays of the sun from the pots and so prevent a too rapid loss of moisture from the soil.

An even better method is to fix a 10 inch board along, and 18 inches from the base of the wall. This will form a trough which is to take the pots, the space around each pot being filled in with peat. This is clean to use and may be kept continually more moist than straw, or boiler ashes may be used. The pots should be placed on a 2–3 inch layer of ashes or peat which may also be placed over the soil of the pots to act as a mulch. An alternative mulch for those living in or near the country, is one of strawy farmyard or stable manure, though this will not be so clean to handle as peat.

Throughout the summer months the roots must be constantly supplied with moisture, a thorough watering being given almost daily, so that the moisture may reach the roots at the very bottom of the pot. The peat or straw around the pots must also be kept moist. To allow the soil in the pots to dry out for only a short period will be to cause irreparable damage for that season.

PLANTING FRUIT TREES

APPLES AND PEARS FOR POTS AND TUBS

Here is a selection of dessert apples and pears suitable for pot or tub culture—

APPLES

Duchess of Oldenburg	August
Lady Sudeley	August
Ellison's Orange	September
Michaelmas Red	September
Egremont Russet	October
Sunset	November
King of the Pippins	December
Adam's Pearmain	December to January
Claygate Pearmain	December to March
May Queen	April to June

PEARS

Laxton's Superb	August
Beurré Bedford	September
Gorham	September
Conference	October
Louise Bonne	November
Glou Morceau	December
Roosevelt	December to January
Winter Nelis	December to February
Santa Claus	December to February
Bergamotte d'Esperen	February to March

Though mentioned elsewhere that Lady Sudeley is a tip bearer, this is correctly so, but it bears its fruit on very short twigs or shoots, and may be said to come somewhere between the tip and spur bearers and is very suitable for pot culture.

PLANTING

A very large pot or small tub should be used, so that the roots are not unduly restricted and may be able to obtain the maximum of food from the compost. Crocks or broken brick should be placed at the bottom of each so that the drainage holes are kept open and over these should be placed a small quantity of fresh turf loam. Do not use the ordinary soil to be found in a town garden, which will generally be sour and completely lacking in nutriment.

Then carefully remove the tap root and trim off any unduly large roots before placing the trees in the pots, spreading out the roots as previously described.

The compost should consist of turf loam to which has been added a small quantity of old mushroom bed manure, or well decayed farmyard manure, but not too much, for an excess of nitrogen must be guarded against otherwise the trees will make too much wood and foliage. But potash is important, $\frac{1}{4}$ oz. of sulphate of potash being allowed for each pot and which must be thoroughly worked into the compost. This should be friable so that it may be carefully packed round the roots and the pot filled to within 1 inch of the rim. The cane is then placed into position and immediately fixed to the wall.

It is not necessary to wait for the ending of the usual winter frosts before planting if the compost is made up indoors (a cellar or shed); planting may be done any time from mid-November until mid-March, but the six weeks preceding Christmas is the best time. This will enable the trees to become thoroughly settled in their new quarters before coming into bloom late in spring.

CULTURE

The care of the trees will be carried out on the same lines as described for all other trees in the artificial form, but help should be given with the setting of the blossom by dusting the individual blooms with a camel hair brush during a dry day, and on several occasions during flowering time. If suitable pollinators are also planted together, there should then be a heavy set of fruit.

Help may also be given the trees to satisfy their moisture requirements by frequent syringing of the foliage, from early June onwards, but if this is done whilst the trees are still in bloom, it must be done in time for the moisture to have dried off before nightfall, as damage might be done by late frosts if the blooms are wet.

The trees will also benefit from feeding once each week with diluted liquid manure water (obtainable in bottles from any sundriesman), from early July when the fruit is beginning to swell. This should be continued until the end of September for the trees will benefit in addition to the fruit.

Where growing in a sheltered position, the fruit may be allowed to hang almost until Christmas, being removed as it is required, and only that of the very late maturing varieties will need to be stored for use in the New Year. This should be removed by the third week of December, when the trees growing in pots should be re-potted in alternate years into a completely freshly made up compost. Trees in tubs, which will contain a larger quantity of compost and provide more nourishment, may be allowed to remain without re-potting for a number of years, if systematically fed and never allowed to suffer lack of moisture. During winter the trees will require no artificial watering, but this may be necessary in April, possibly following a long period of frost and drying winds.

CHAPTER XIX

FRUIT TREES THAT ARE ORNAMENTAL

Apple blossom of rich colouring—The richness of pear foliage—Cherry blossom.

Do we appreciate the ornamental value of many of our apple and pear trees as we should? Few will disagree that there is no fruit quite like that grown in an English garden, but whilst much care is given in selecting suitable varieties for their fruit, little thought is given to the exquisite blossoms of certain apple trees, or to the richly coloured foliage of pears. We spend hours in the selection of a particular shrub, which may not possess anything like the attractive colourings of some fruit trees and which in addition, do not bear a crop of delicious fruit.

In the already established orchard at my home, care was taken when cutting out a number of old and cankered trees to replace with a number of those renowned for their ornamental qualities, in addition to the quality and flavour of their fruit.

APPLE BLOSSOM OF RICH COLOURING

The apple being the last of all fruit trees to flower, it is only rarely that the blossom suffers from severe frost damage, and if there is a lovelier sight than a West Country orchard in its early summer glory, I have yet to see it. From the upstairs window of my cottage it is possible to look down on to a sea of all the shades of pink imaginable, only the yellow and white of the daffodils and splash of the rich navy-blue of the grape hyacinths flowering between the tree are noticeable amidst the pink and white blossom of which none is more striking than the rich deep pink of Brownlee's Russet. But this apple is better suited to a cold district and a clay soil, which brings out its aromatic flavour in the same way as the colder regions accentuate the flavour of most gooseberries. Another russet of exceptional keeping qualities is Merton Russet, and though with us for only a short time, its large rich carmine-pink blossom is particularly attractive. But of even greater beauty is the blossom of Arthur Turner, which opens out with exquisite pink colouring. The petals lie flat thus presenting themselves to the fullness of their beauty. The blossom appears before any leaf has formed and it remains fresh over a longer period than any apple with the exception of Lane's Prince Albert, the colour being enhanced by the long deep yellow stamens. The leaves too possess beauty, being the darkest green coloured of all apples. An excellent pollinator for Bramley's Seedling and a more reliable cropper, it makes a sizeable fruit by late July.

Another with most attractive blossom is Upton Pyne, its blossoms being even larger than those of Arthur Turner, and of a rich creamy-pink colour, shaded with crimson, also colourful over a long period.

A very late dessert apple of Canadian origin is Ontario, which grows well in a clay soil and bears an attractive flat fruit of vivid green, striped with bright red and which will keep until May, the flesh remaining white and possessing a pleasant not too sweet flavour. The blossom is most beautiful, being ivory coloured, edged with pale pink and which appears before the leaves.

Another with particularly lovely blossom is Annie Elizabeth, one of the best of all long-keeping cookers, its fruit remaining sound until the first of the Emneth Early fruit is ready in early July. Flowering late it is valuable for a frosty garden, the blossoms being of a deep carmine-pink, heavily veined deep crimson.

As a contrast plant with it Woolbrook Pippin, whose blossoms are remarkable for an apple tree in that they are of the same grey-white colour as those of the pear.

A cooking apple of value which has been rather neglected is Cottenham Seedling, of Yorkshire origin and which crops really well in that county. It is almost as pleasant when cooked as Bramley's Seedling, and it will keep if carefully stored until Easter. It bears a huge yellowish-green apple, streaked with russet and red. The blossom is white, flushed with rose-pink, and it is late to appear, at the same time as the leaves.

The last of all apples to bloom is Crawley Beauty, a long-keeping dual-purpose apple of quality. This is a splendid apple for growing in frosty areas for the pink and white blossom does not appear until early June, which also lends colour to the orchard when all other blossom has ended.

The flowering crab apples too should not be neglected. They are delightful planted down a drive or in the shrubbery. One of the most free flowering is Pyrus floribunda, which covers itself in a cloud of pink blossom during April. Another of compact habit is P. Eleyi, the long arching branches, purple tinted, are covered with rich crimson flowers in spring, whilst the fruit hangs like dark red cherries during autumn. To make the best jelly are the crimson and yellow fruits of P. Dartmouth, which bears pure white blossom in April. These crabs make a colourful background to an orchard and are useful as a windbreak.

RICHNESS OF PEAR FOLIAGE

Most pear blossom is pure white, the dark brown anthers providing a striking contrast. The blossom blends with the first crimson and pink of the early flowering apples, making a most pleasing picture in the mixed orchard, but it is in the early autumn that the pear tree is most ornamental, the brilliant autumnal leaf colourings and the attractively tinted fruits providing a display of great richness. No varieties of any species of tree produce such varied colourings, from the bright yellow foliage of the vigorous Jargonelle, described by Parkinson in his *Paradisus* (1629) as being, " of a very pleasant taste," to the deep crimson leaves of the russet-red fruited Santa Claus, one of the most richly flavoured of all pears. Also bearing crimson foliage in autumn is Pitmaston Duchess, its large pale yellow fruit providing a marked contrast.

One of the most fertile of all pears and producing its juicy aromatic fruit early in October is Beurré Hardy, its autumnal leaf colouring being of richest scarlet similar in colour to the foliage of Beurré Clairegeau, which also bears a large fruit of pale yellow, strikingly marked with scarlet. In the cold Northerly districts the fruit rarely seems to ripen as it does in the South and though not of first rate dessert qualities, is a good culinary pear. Another pear which will keep probably longer than any other is the little known Josephine de Malines. It bears only a small fruit but retains its delicious nutty eating quality until the end of winter, the fruit carrying the unusual delicate perfume of orange blossom. The foliage turns a vivid primrose-yellow colour in autumn which remains on the tree for a considerable time.

Another pear which is rarely seen in gardens today is Beurré Bosc, which bears a long tapering fruit of a deep golden, brown colour, and is highly aromatic. Its foliage takes on the most attractive range of autumn tints, which together with the richly coloured fruit makes a glorious display in the orchard.

CHERRY BLOSSOM

Taking at least ten years to come into reasonable bearing cherries are now little planted, which is a pity, for if only for the magnificence of their blossom in late spring at least two or three trees should be in every garden. The best variety is still the early April blossoming Early Rivers, which bears its jet black fruits early in June. It requires a pollinator, the best being the early flowering and late maturing Bradbourne Black, and for midseason which is also pollinated by the others, is the new Merton Heart.

In a shrubbery or about the orchard or even used as a windbreak, damsons will provide valuable fruit in autumn and will enhance the garden in spring with its feathery-white blossom there being no more arresting sight amongst flowering trees. The two best varieties for providing colour in spring and late autumn are Merryweather and Westmorland, and which retain the purple-black fruit on the trees long after the leaves have fallen.

From the pink blossom of the cherry, and the crimson and white shades of apple blossom in spring, to the rich autumnal colourings of the foliage of the pear tree, the orchard planted for its ornamental value as well as for its eating qualities will provide eight months of continuous interest and splendour, unequalled by any other ornamental tree.

PART II

THE STONE FRUITS

CHAPTER XX

THEIR TRAINING AND PRUNING

Plums—Spring Pruning—Removal of suckers—Treatment of the fan trained tree—Forming the fan shaped tree—Sweet Cherries—Morello Cherries.

PLUMS

SPRING IS the best time to carry out any pruning of plums, just when the buds are beginning to burst for it is at this time that the wounds quickly heal over and almost no 'bleeding' occurs. This not only reduces the vigour of the tree but provides an entrance for the dreaded Silver Leaf disease, the fungus deriving its nourishment from the cells of the tree, thereby greatly decreasing its constitution. Early autumn pruning, which may be carried out on early fruiting varieties when the crop has been cleared, is permissible, but all cutting should be done between the end of April and mid-September, for during the winter the cuts will remain 'open' for dangerously long periods.

In any case, plums require very little pruning, for the trees will form their fruit buds throughout the whole length of the younger branches, and especially with standard trees which are established; thinning of overcrowded growth either in May or September, depending upon lateness of crop, will be all that is required. A well-grown plum tree will be able to carry a much larger proportion of wood than will any other fruit tree and drastic reduction, even of neglected trees, must never be performed as with apples and to a lesser extent, pears.

When renovating a neglected tree, it may be advisable to cut away with the pruning saw one or two large and partially decayed branches. If so, this should be done during May, a time when the large cut will heal rapidly and so that the energies of the tree may be concentrated to the remaining wood. With plums it is even more important to cut out any wood close to the stem from which it is being removed so that the wound will heal rapidly and completely. But before making any cuts, see if the tree can be renewed in vigour by removing some of the small, thin wood which plums make in quantity, and possibly root pruning will be more satisfactory than the cutting back of any large branches.

REMOVAL OF SUCKERS

One of the greatest troubles with plums is the continual formation of suckers at the roots which if left will untilise much of the nourishment needed for the proper functioning of the tree. These should be removed whenever the roots of the tree are pruned, and must be cut away right from their source otherwise they will continue to grow again. It is first necessary to remove the soil from around the tree to expose the roots, but it will be found that the suckers generally arise from a point in the roots just below the point where the scion has been grafted on to the rootstock. This calls for the utmost care in removing the soil right up to the scion and then in cutting out the sucker shoot with a sharp knife. For bush or standard trees it is advisable to ring round half the tree one year and to complete the removal of suckers and vigorous roots the following year. It is essential to pack the soil well round the roots when the work has been done or there will be the chance of the tree becoming uprooted by strong winds.

TREATMENT OF FAN TRAINED TREES

In renovating fan trained trees, in which form the plum crops abundantly, more pruning will be necessary and this should take the form of pinching back shoots in mid-summer and in removing completely all unwanted new wood. A number of young growths may be pinched back between mid-June and mid-July to form a new spur system and these will need to be cut further back, in the same way as described for apples, but early in September rather than in winter. Then by degrees the old spurs may be drastically reduced to make way for the new ones. In conjunction with the shoot thinning of wall trees, root pruning should be given every three or four years which will prevent excessive wood growth.

The formation of a bush and standard form of plum tree takes the same lines as described for apples and pears. Planted in the maiden form they may be formed as required, the yearly pruning consisting of pinching back the new wood to form fruiting buds.

FORMING THE FAN TRAINED TREE

Both plums and cherries crop abundantly in this form, all varieties proving suitable, though naturally some are more vigorous than others and will require more frequent pruning at the roots. The method of forming the fan is to cut back the maiden to an upward bud. This should leave on the lower portion of the stem two buds, which will break and form the arms. Unsuitably placed buds should be removed, and any not breaking must be nicked or notched as previously described for the formation of espaliers.

After the previous season's growth, they are pruned back to 18 inches and the leader or central shoot is cut back to two buds. It is from these buds that the fan shaped tree is formed.

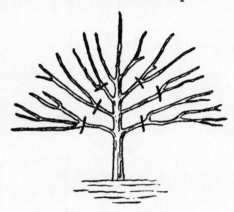

Forming the fan tree.

Canes are used for tying in the shoots so that they may be trained to the required shape. As growth continues, each shoot may be cut back the following spring to two more buds which will complete the shape of the tree, though canes will be needed until the shoots have taken on the required form.

Cultural treatment will henceforth consist of cutting back a third of the new wood formed by the branches each May, and the pinching back of all side growths. The shoots will continue to break and where there is room a number may be tied in to continue the fan-like shape.

SWEET CHERRIES

Being much slower growing, the cherry will make considerably less growth than the plum and so will need less pruning. This is all to the good for the serious 'bleeding' or gumming of the cherry where cut, will sap its energy and also like the plum will be a source for Silver Leaf infection. August seems to be the best time for any pruning, after the crop has been gathered. All dead, or those branches interfering with each other, should be cut away and where any dense growth has been made, which may not be until the tree is 20 years old, this should be thinned out.

In the case of the fan-trained tree, it is only the side growth which should be removed, pinching back to about six leaves late in June and then further pinching back to four buds early in September. If it is noticed that the main leaders are making excessive growth, rather than cut them and cause further stimulation, they will respond better to the bending down of the shoots for a period of twelve months bringing them forward from the wall and tying the tips against the main stem. This will weaken growth and they may be pruned and tied back in their normal position in 12–18 months. Root pruning as described will do much to keep the cherry in check.

MORELLO CHERRIES

The Morello or acid cherry requires a different treatment. Acid or sour cherries, bear their fruit on the previous season's wood, and so the aim must be to encourage a continual supply of new wood. For this, the laterals should be cut back halfway each autumn, then in spring all side growths should be cut or pinched back to two inches of growth. The previous season's shoots should now be bearing both new wood and blossom buds and it is upon these that the present year's fruit is borne. At the base will be found several buds which should be removed to leave but a single bud which will continue to make new growth through the summer for fruiting next season. Like the sweet cherries and plums, the Morellos can be allowed to carry plenty of wood without fear of reducing the amount of fruit, so thinning is not so necessary and should be done only when need arises. Root pruning will help to keep wall trees in check. As the acid cherries are prone to Brown Rot Disease, all diseased wood when observed must be carefully removed.

CHAPTER XXI

PLUMS AND GAGES

i. THEIR CULTURE

The value of Plums in the garden—Growing against a wall
Pollination—Soil conditions—Propagation and rootstocks.

NO FRUIT is quite so rich and delicious as the plum, especially those possessing Green Gage blood, having a juicy, almost treacle-like sweetness, and a long period of maturity from late July until November, when several varieties will still be hanging in the trees, whilst others may be stored until that time and possibly longer. For a northern garden, or one troubled by frosts, then it will be wiser to plant late flowering apples, damsons, and the very latest of the plums to bloom, for as a general rule, the plum is the first of all the fruits to open its bloom, at the beginning of April. For this reason, plums planted commercially are always the most unreliable of all fruits, for they either escape frosts, and crop so abundantly that the price of fresh fruit is uneconomical to the grower; or they may be badly damaged by frosts, when there is little fruit available. Where frosts prove troublesome, such as where the land is low lying, or perhaps close to a river, then all but the very latest flowering plums should be avoided, and these should be given the protection of a wall. Plums do not require such a large amount of sunshine as pears, and provided they may be given frost protection, they will ripen well on a west or east wall, and so leave the southerly positions for the pears, with the apples being planted in the more open and exposed parts of the garden.

GROWING ARAINST A WALL

Plums do better as bush trees than as standards, and in the fan-shaped form rather than the horizontal form which suits the pear best. Where planting in a small garden, apples should be planted as cordons, pears as espaliers, and plums as fan-shaped trees against a wall. This is not only the most reliable method, but the most economical, for a wall plum, well supplied with moisture and nitrogen, will soon cover an area of 10 ft. high and a similar distance in width.

Suitable varieties will depend upon the aspect and district, the plums being more hardy than the gages, whilst there are also plums more hardy than others, and the selection should be made accordingly, planting the late flowering plums and gages in a more open situation, the more tender and choice dessert varieties against a wall. So many walls are clothed in uninteresting ivy, when they might be growing delicious fruit, but with wall trees it must be remembered that they must never lack moisture. Lack of humus and moisture at the roots is the cause of so many giving disappointing crops. A wall, especially where it receives some sun will bring out the flavour to its maximum, but if the trees are not supplied with abundant summer moisture, the fruit will remain small, the flesh dry.

The plum will come quickly into bearing in the fan-form, and is in no way troubled by biennial bearing as are many apples, neither does it require the same attention to pruning as either the apple or pear. It bears the bulk of its fruit on the new wood, and apart from the removal of any dead wood, the pruners are better left in the garden shed. Excessive pruning, for pruning's sake, and especially in the dormant period, will cause untold harm by ' bleeding,' from which all stone fruits suffer. The Government's Silver Leaf Disease Order, demands that all dead and decayed wood is removed and burnt by July 15th each year, and this, together with any shortening of unduly long shoots, should be done early in spring, just when the trees are coming into life after their winter rest. At this time any cuts will quickly heal over.

POLLINATION

As with apples, usually the most richly flavoured plums and gages require a pollinator, especially where the connoisseur's plum, Coe's Golden Drop has been used as a parent, for its blossoms are sterile and will not pollinate each other. About half the most popular plums possess self-sterile blossom and require a pollinator, whilst the rest are self-fertile and extremely so and will, unlike apples and pears, bear a heavy crop entirely without any pollinator. A number however are only partly self-fertile, and will set heavier crops with a pollinator flowering at the same time. Again, plum pollination is less complicated, in that the blossom period of the entire range of plums covers only about 18–19 days, and except for the very latest to bloom such as Marjorie's Seedling and Pond's

Seedling, most will overlap, and so the flowering periods may be classed as Early and Late for pollinating purposes. It may be said that the early flowering self-fertile varieties will pollinate the early flowering sterile varieties, and the same with those which bloom late. It is therefore a more simple matter than when considering apples, pears, and cherries, where many varieties do not prove suitable pollinators, though in bloom at the same time. Of all plums only the following in no way overlap, or only by a day or so, and those of group (a) should not be relied upon to pollinate those of (b) and vice versa—

Early (a)	*Late* (b)
Bryanston Gage	Belle de Louvain
Count Althann's Gage	Czar
Jefferson	Late Transparent Gage
Monarch	Marjorie's Seedling
President	Oullin's Gage
Warwickshire Drooper	Pershore

All plums and gages remain in bloom, unless damaged by frost, for exactly 10 days, a shorter period than any other fruit, and those of group (a), the first to bloom, will have almost finished when those of group (b), the last to bloom, commence to flower.

It should be said that President is incompatible as a pollinator to Cambridge Gage, though both are in bloom at the same time.

Unfortunately many gardeners after giving careful consideration to the selection of suitable varieties for various soils and climates, do not take the pollinator factor into consideration, it being pure luck if the self-fertile varieties are planted, and where self-sterile varieties fail to set their fruit, this is blamed on the soil, climate, or even upon the nurseryman.

SOIL CONDITIONS

Strange as it may seem that whereas by far the greater number of dessert apples and pears having during this century been planted in Britain as against those for culinary use, the opposite is the case with plums, the greater percentage such as Czar, Yellow Pershore, and Belle de Louvain, being planted for the canning and jam industries. The result is that if we wish to enjoy those richer flavoured plums and gages, we must grow them ourselves, but before making a selection, soil must be considered with some care.

Plums like a heavy loam, and John Scott in his *Orchardist*, the commercial growers' bible, published about 1860, says " plums succeed best in strong, clay soils, mixed with a proportion of loam. On such soils the plum reaches the highest perfection in the shortest possible time." A soil retentive of moisture is the secret of success with plums, and one continually enriched with nitrogen, preferably of an organic nature.

Light soils should have large quantities of shoddy, or strawy farmyard manure incorporated at planting time, and especially where planting wall trees. A liberal mulch of an organic manure, rich in nitrogen, should always be given in April each year. Where this is unobtainable, give 1 oz. of sulphate of ammonia to each tree at the same time, and immediately after any pruning has been done.

Especially with wall trees and with recently planted trees in the open, artificial watering should be done whenever necessary during a dry summer.

Though several varieties, such as Pond's Seedling and the old Green Gage, will crop well and remain healthy on a chalk laden soil, most other varieties soon show, like pears, signs of chlorosis and never bear well. When there is a reasonable depth to the soil this may be largely overcome by working in plenty of humus forming manure. Under opposite conditions, where there is almost a complete absence of lime in the soil, the two most tolerant varieties are Czar and Victoria, for as long as the soil is heavy and not waterlogged, they will bear well. Shoddy and composted straw should be added where the soil proves excessively heavy and sticky. Czar, possibly the most accommodating of all fruit trees, will also bear abundantly in a light, sandy soil containing a very small percentage of loam.

One may ask what weight of fruit is to be expected from a 20-year bush tree, growing in an average soil, and being constantly mulched with organic nitrogenous manures? Taking good years with the bad, and a wide selection, the average should be about 40 lbs. per tree, with Victoria, Czar and Pershore, as high as 50–60 lbs., and with Coe's Golden Drop and Kirke's Blue as low as 8–10 lbs. per tree, but still worth growing for their delicious fruit. It should be said however, that where given favourable conditions, the shy bearers will crop more heavily, whereas there may not be a marked difference with the others, for instance, when planted against a sunny wall in a southern garden, with their roots well supplied with moisture and nitrogen.

PROPAGATION AND ROOTSTOCKS

The choice of rootstock is not large, and generally bush and fan-trained trees are grown on what is known as the Common Plum and Brompton Stock, and standard trees for large gardens or orchards on the Myrobalan stock. The latter is generally used for the heavy cropping varieties, such as Czar and Monarch. Owing to the incidence of gumming and the chances of introducing disease, plums are budded rather than grafted. Budding is done in July, for plums and cherries, the bud being removed with a strip of bark, cut out with the pruning knife. This is then fixed against the wood of the selected stock, into which a cut in the bark has been made 6 inches from the base. The bark on both sides of the cut is carefully lifted from the cambium layer, and into this the bud is fixed. It is held in position by tying with raffia, leaving only the actual bud exposed. In from 4–5 weeks the union should have taken place, when the raffia is removed. The following March the stock is cut back to within 1 inch of the bud, which will grow away to form the tree.

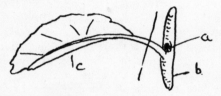

A bud for propagation. (a, bud at leaf joint; b, bark cut from tree; c, leaf [which is removed where shown])

The gages and plums in the artificial forms are mostly budded on to the Common Plum or Brompton stocks. The varieties President, Czar and Marjorie's Seedling however are incompatible with the Common Plum stock and so are usually budded onto the Myrobalan. The Common Plum makes a robust tree, is resistant to Silver Leaf, and so is always used for Victoria and Pond's Seedling, whilst it is generally used for the less vigorous varieties such as Coe's Golden Drop.

For a small garden, the Brompton Stock is probably the best, for the trees grow sturdily, but come quickly into bearing, whilst they send up few suckers, an important consideration for the amateur, requiring but little attention apart from the occasional removal in spring of any dead wood.

At one time the Common Mussel stock was widely used, but the trees on this stock require copious amounts of water and tend to sucker badly. With this stock the trees come more quickly into bearing than on any other, but like Type MIX with apples, bear abundantly for a time, then lose vigour.

It should be said that several of the gages, especially Oullin's Golden and Count Althann's, are incompatible with the Myrobalan stock; whilst several others, such as Yellow Pershore, crop well and are generally planted on their own roots, but being slow to form suckers they cannot be used for propagating other plums on a commercial scale.

CHAPTER XXII

PLUMS AND GAGES

ii. VARIETIES

PLUMS

RIPE LATE JULY AND EARLY AUGUST

Black Prince. Ripe before the last days of July, it makes a small tree, yet is a huge cropper, the small black velvety fruit having the true damson flavour, and being delicious for tarts and for bottling. It is extremely resistant to Silver Leaf Disease.

Blue Tit. Raised by Messrs. Laxton Bros., and follows Czar. It bears a blue fruit with the true greengage flavour, and is one of the best early plums for dessert. Makes a small, compact tree and is very fertile.

Czar. Makes a fine orchard tree with its vigorous habit, yet it is compact and a most reliable cropper. Its blossom appears late and is very resistant to frost, and so it bears heavily in cold gardens, and especially in heavy soils. The fruit is of a bright shade of purple, of medium size, and is useful both for cooking and for dessert. Introduced and raised by Thomas Rivers in 1875, it is ready for use at the beginning of August.

Early Laxton. The first plum to ripen, towards the end of July. The small, golden yellow fruit carries a rosy-red flush, and is sweet and juicy. It blooms early and is pollinated by River's Early Prolific or Laxton's Cropper. It makes a small tree, and is valuable for a small garden.

River's Early Prolific. Making a small, but spreading tree, this is a good companion to Early Laxton, for it is grown chiefly for cooking, the small purple fruit possessing a rich damson-like flavour. Pollinated by the early flowering gage, Denniston's Superb. Hogg says, " rarely ever misses a crop."

RIPE MID-AUGUST TO MONTH END

Goldfinch. To ripen mid-August this is possibly the best of all plums. Raised by Laxton Bros., it has Early Transparent Gage blood, and is equally as delicious, the golden-yellow fruit being sweet and juicy. It makes a compact tree, and bears consistently heavy crops.

Laxton's Bountiful. Has Victoria as a parent, and bears a similar fruit, but of not quite so good flavour. It bottles better than any early plum, and should also be used for jam rather than for dessert. It makes a large, vigorous tree and bears enormous crops.

Pershore (Yellow). Used almost entirely for canning, bottling and jam, it was found in a Worcestershire garden, and is widely grown in that county. It makes a bright yellow fruit with firm flesh. It is valuable for a frosty garden, for it is very late flowering.

Utility. One of the most handsome plums and a fine all-round variety, raised by Laxton Bros., and introduced forty years ago. It bears a large exhibition plum of bright purple-red. Early flowering it may be pollinated by most of the early blooming plums, especially Denniston's Superb. Matures between Goldfinch and Victoria.

Victoria. Found in a Sussex cottage garden more than a century ago, and the most widely grown plum of all. Extremely vigorous, it is the most self-fertile of all plums; it is frost resistant, crops well in all forms, and is used for every purpose. Its only weak point is that it is often troubled by Silver Leaf Disease. Ripe at the end of August. Like Czar, it crops well in clay soils.

RIPE EARLY TO MID-SEPTEMBER

Angelina Burdett. Known to early 18th century gardeners and much too good to become extinct, yet I know of only one nursery where it is still propagated. The large fruit is of deep purple, speckled with brown. It is ripe at the very beginning of September, but will hang for a fortnight when, as Hogg says, " it forms a perfect sweetmeat." It will also keep for a fortnight after removing, and as it is extremely hardy, and is a regular bearer, it should surely be in every garden.

Belle de Louvain. Of Belgian origin, it makes a large tree, and is slow to come into bearing, but for bottling and cooking, it is most valuable, also for a frosty garden on account of its late blooming.

Giant Prune. Raised by the famous Luther Burbank of California, from Pond's Seedling, to which it is similar in size, colour and flavour. It also blooms late and is valuable for a northern garden.

Jefferson. Almost like a gage in its flavour and rich dessert quality, the pale green flesh being sweet and juicy. Raised in the U.S.A. it blooms early and requires an early flowering pollinator such as Denniston's Superb. It makes a compact, upright tree, ideal for a small garden, but should be planted in the more favourable districts. The fruit is pale green, flushed with pink.

Kirke's Blue. To follow immediately after Jefferson, this is an equally delicious plum for dessert, its large violet fruits, being sweet and juicy. Introduced by Joseph Kirke of the Old Brompton Road, London, about 1825, it is a shy bearer in the north. Czar and Marjorie's Seedling are the two best pollinators.

Laxton's Cropper. This is an excellent all-purpose plum for September, the large black fruit hanging for several weeks when ripe. It is a strong grower and bears a heavy crop in all districts. Will store well.

Pond's Seedling. It makes a large tree and is valuable for cold gardens, in that it blooms late. The rose-crimson fruit is large and handsome, and if not of the very best dessert qualities, it is good. Crops well in a chalk laden soil.

Thames Cross. A new plum raised at Long Ashton. It bears abundantly in the West Country, where it makes a large tree, and bears large pure golden-yellow fruit. Has Coe's Golden Drop as a parent, and the flavour is similar.

Warwickshire Drooper. Making a large, vigorous tree with drooping branches, this is an excellent all-purpose plum, where room is available. May be described as a later and improved Pershore. The yellow fruit is shaded with scarlet and grey.

White Magnum Bonum. Hogg describes it as a "culinary variety, highly esteemed for preserving." It blooms very late and is valuable in this respect, whilst the pale yellow fruit is borne in abundance. It makes a huge, spreading tree, and does well in a heavy soil.

RIPE LATE SEPTEMBER AND EARLY OCTOBER

Coe's Golden Drop. It blooms very early and should be given the protection of a warm wall, whilst it also likes a soil containing plenty of nitrogenous humus. It bears a large fruit of quite exceptional flavour, pale yellow, speckled with crimson. At its best about October 1st, the fruit will keep until the month end. Its rich apricot flavour, and almost treacle sweetness, being the most delicious of all plums. It makes a spreading tree, requires an early flowering pollinator, and is a shy bearer. Raised 200 years ago by Coe of Bury St. Edmunds, this is the Cox's Orange or Comice of the plum world. Denniston's Superb is the best pollinator.

Laxton's Delicious. This is one of the finest of dessert plums, the deep yellow fruit, flushed with red being juicy and deliciously sweet. It has Coe's Golden Drop as a parent, but is a much better cropper, especially as a wall tree. Like its parent the fruit may be kept several weeks, if harvested about the 3rd week of September. A vigorous grower, it blooms late, Oullin's Golden Gage or Marjorie's Seedling being suitable pollinators.

Laxton's Olympia. Making a large, spreading tree, it blooms late and ripens its fruit about the 1st October. The coal black fruit is of medium size, is sweet, and possesses a flavour all its own, similar to preserved plums from the Mediterranean. It bears a very heavy crop.

Marjorie's Seedling. Extremely fertile and of vigorous upright habit, it makes a large tree. It is the last to come into bloom, and the latest to ripen its large crimson-purple fruit. Ready for gathering at the end of September, the fruit will hang until the end of October, when it may be used for all purposes. Raised in Staffordshire, it is most suitable for a cold northerly garden.

Monarch. Similar in all respects to Marjorie's Seedling, the tree habit and quality and colour of the fruit being the same, it must be considered inferior to Marjorie's Seedling, in that it blooms very early, and is frequently damaged by frost, neither does the fruit hang so well.

President. Raised by Thomas Rivers and an excellent dessert plum, being large, rich purple, with its deep yellow flesh juicy and sweet. Makes a large, spreading tree. It blooms early and requires a pollinator.

Severn Cross. Raised by Mr. Spinks at Long Ashton, this is the latest of all dessert plums, for it hangs well into October, and is valuable where a succession of fruit is required. It makes a tall, vigorous tree, the fruit being golden-yellow, flushed and spotted with pink, extremely juicy and of good flavour.

Extremely hardy plums—
    Angelina Burdett
    Czar
    Early Prolific
    Laxton's Delicious
    Pershore
    Pond's Seedling
    Victoria

Long hanging plums—
    Angelina Burdett
    Laxton's Delicious
    Marjorie's Seedling
    Pond's Seedling
    Severn Cross

Plums of vigorous, spreading habit—
    Coe's Golden Drop
    Czar
    Laxton's Bountiful
    Laxton's Olympia
    President
    Warwickshire Drooper
    White Magnum Bonum

Plums of dwarf, compact habit—
  Black Prince   Jefferson
  Early Laxton   Kirke's Blue
  Goldfinch

With plums, pollination may be divided into two sections, those that bloom early to mid-season, and those in bloom mid-season to late—

        S.S. = Self-Sterile;   S.F. = Self-Fertile
Early Flowering
Black Prince   (S.S.)   Monarch   (S.F.)
Blue Tit   (S.F.)   President   (S.S.)
Coe's Golden Drop   (S.S.)   Thames Cross   (S.F.)
Early Laxton   (S.S.)   Victoria   (S.F.)
Early Prolific   (P.S.F.)   Warwickshire Drooper   (S.F.)
Jefferson   (S.S.)

Late Flowering
Angelina Burdett   (S.F.)   Marjorie's Seedling   (S.F.)
Belle de Louvain   (S.F.)   Pershore   (S.F.)
Czar   (S.F.)   Pond's Seedling   (S.S.)
Giant Prune   (S.F.)   Severn Cross   (P.S.F.)
Kirke's Blue   (S.S.)   White Magnum Bonum   (S.F.)
Laxton's Delicious   (S.S.)

Though the self-fertile varieties will set fruit with their own pollen, they will set much heavier crops when planted with varieties in bloom at the same period.

GAGES

Outstanding with their own particular flavour, the gages as a general rule are less hardy than most plums, and should be given a sheltered, sunny position for them to bear consistently heavy crops. They bear more heavily planted south of a line from Chester to Lincoln.

TO RIPEN EARLY—MID-AUGUST

Denniston's Superb. Really a gage-plum hybrid, but possesses the true gage flavour, and is extremely hardy and fertile. Like James Grieve amongst apples, this plum acts as a pollinator for more plums than any other variety. It was raised in New York in 1835, it blooms early mid-season, and can set heavy crops without a pollinator. It is of vigorous habit in all soils, and ripens its green fruit, flushed with crimson by mid-August. One of the best of all plums or gages, whichever it may be considered.

Early Gage. Raised by Laxton Bros., this is the first of the gages to ripen, at the beginning of August. The tree is vigorous and healthy, and when pollinated (Denniston's Superb) bears heavily, the amber-yellow fruit possessing a rich, but delicate flavour.

Early Transparent. Raised by Thomas Rivers, it makes a dwarf tree and is able to set a heavy crop with its own pollen. It blooms early and ripens its fruit during mid-August, when the pale apricot skin is so thin as to show the stone. The richly flavoured fruit possesses a distinct fragrance when ripe.

Oullin's Golden Gage. " Ripe mid-August and a remarkably fine dessert variety," wrote Dr. Hogg. Like Denniston's Superb, this seems to be a hybrid raised in France a century ago. It is valuable in that it is one of the latest gages to bloom, and though good for dessert, it is one of the best for bottling and jam.

RIPE LATE AUGUST—EARLY SEPTEMBER

Cambridge Gage. Raised and used by Chivers of Cambridge, it bears a fruit similar to the true Green Gage, but is hardier and is a heavier cropper. Flowering late, it is a valuable variety for the North Midlands.

Late Transparent. Making a small, dwarf tree, and setting a heavy crop with its own pollen, it possesses similar characteristics to Early Transparent, though it blooms later, Laxton's Gage being a pollinator. The bright yellow fruit is speckled with red, the flavour being rich, almost peach-like.

Laxton's Gage. The result of Greengage x Victoria, it makes a large, spreading tree and blooms quite late. It is a useful variety for Midland gardens. The yellowish-green fruit, which possesses a rich flavour is ripe at the end of August. It is a heavy bearer in most soils.

RIPE MID-SEPTEMBER TO EARLY OCTOBER

Bryanston. The fruit is ripe mid-September, being pale green, speckled with crimson, and with a rusetting nearest the sun. It makes a large, spreading tree, is early flowering, and with Victoria as a pollinator, which is essential, it crops profusely. The late Edward Bunyard says in *The Anatomy of Dessert,* "its flavour quite equals the Green Gage . . . it is yet too little known," though found in a Dorsetshire garden at the beginning of the 19th Century.

Count Althann's Gage. One of the most richly flavoured of all the gages, it makes a large, but compact tree, blooms late and ripens its fruit towards the end of September. Introduced from Belgium a century ago, the fruit is unusual for a gage in that it is dark crimson, speckled with brown. Should be eaten as soon as ripe. Must have a pollinator when it crops heavily.

Golden Transparent. Like all the Transparents, raised by Thomas Rivers, it makes a dwarf tree, is self-fertile and blooms late. The fruit, which is possibly the most delicious of all gages or plums, ripens early October, the last of the gages to mature.

Greengage. At its best during September, depending upon the locality, when its greenish-yellow fruit is rich and melting, and faintly aromatic. Known to early 18th Century gardeners, it is a shy bearer unless pollinated with Victoria, and should be grown in a sheltered garden. It was introduced from France in 1720 by Sir William Gage, and grown in the garden of Hengrave Hall, Suffolk.

Reine Claude de Bavay. Hogg describes it as a " first rate plum of exquisite flavour." It is ripe about October 1st, and will hang for several weeks. It makes a neat, compact tree, and is self-fertile, blooming very late. The richly flavoured fruit is large and almost orange in colour, speckled with white. The best gage for a small garden.

Late Flowering—
Cambridge Gage	(S.S.)	Laxton's Gage	(S.F.)
Count Althann's Gage	(S.S.)	Oullin's Golden Gage	(S.F.)
Golden Transparent	(S.F.)	Reine Claude de Bavay	(S.F.)
Late Transparent	(S.F.)		

Early Flowering—
Bryanston	(S.S.)	Early Transparent	(S.F.)
Denniston's Superb	(S.F.)	Greengage	(S.S.)
Early Gage	(S.S.)		

Very late Flowering—
Belle de Louvain	(S.S.)	Marjorie's Seedling	(S.F.)
Czar	(S.F.)	Oullin's Golden Gage	(S.S.)
Late Transparent Gage	(S.S.)	Pond's Seedling	(S.F.)

Gages of dwarf, compact habit—
Early Transparent   Late Transparent
Golden Transparent   Oullin's Golden Gage
Greengage   Reine Claude de Bavay

Most suitable varieties for a chalk soil—
Greengage   Pond's Seedling
Marjorie's Seedling   River's Early Prolific

CHAPTER XXIII

DAMSONS AND BULLACES

WITH THE exception of the counties of North-West England, from the borders of Shropshire and Cheshire, to the borders of Scotland, where grown on a considerable scale, the Damson, a native of the country around Damascus, hence its name, remains comparatively neglected in Britain. Yet it is so hardy that it could well be used much more for providing a shelter or windbreak for fruit trees in an exposed district. Or for that matter, these hardy fruits may be planted as a substitute for the earlier flowering plums. As to soil, they will crop abundantly in a thin soil, provided they are planted in an area of excessive rainfall, such as the Western side of Britain, for damsons flourish in abundant moisture, as do plums. They should also be given a nitrogenous dressing each year, preferably in spring. Damsons bloom later than plums and the strongest of cold winds do not trouble them. Retaining their foliage right through autumn, they provide valuable protection for other fruit trees.

Possessing a flavour and fragrance all their own, delicious used for tarts and pies, and for making jam, also bottling to perfection and retaining their flavour for several years if necessary, the damsons are one of the most valuable fruits for an exposed garden, yet are comparatively neglected today, and the modern generation, as with Claygate Pearmain apple, does not realise what it is missing.

With the exception of Farleign Prolific, all will set fruit with their own pollen, but as with most fruits, where two or more varieties, possibly for succession, are planted together, heavier crops result.

These hardy fruits may be planted about the garden where others would not grow well, or they may be planted in a hedgerow, or as a shelter belt, their silver-grey blossom being most ornamental, and their fruit most attractive throughout autumn.

DAMSONS

Bradley's King. The best variety for a Northern garden, for it is extremely hardy, blooms late and is a heavy bearer, besides making a vigorous tree, the wood not being brittle. It bears its fruit mid-September and is almost as large and richly flavoured as the Shropshire Prune, being of an attractive dark crimson colour, whilst the foliage takes on the autumnal tints usually associated with the pear.

Farleigh Prolific. Found growing in Kent at the beginning of the 19th Century, this is the most prolific bearer of all, if given a pollinator, e.g. Bradley's King. Known also as Crittenden's or The Cluster Damson, its fruit hanging in huge clusters, it is the first of the autumn damsons to mature, ready for use early September. Its small coal black tapering fruits make superb jam. It forms a small compact tree and is generally planted in South-East England and East Anglia.

Merryweather. With its large, round blue-black fruit, it may easily be mistaken for a plum, yet it possesses the rich flavour and fragrance of a true damson. It makes a large, spreading tree and blooms quite early, so should be planted where late frosts are not troublesome. Yet it is extremely hardy and bears a heavy crop, which will hang through October. Introduced by H. Merryweather & Sons of Southwell, Notts., of Bramley's Seedling fame.

River's Early. This is the only summer fruiting damson, ready for use early August. It blooms very early and should not be planted where frosts prove troublesome. In more favourable gardens it sets a heavy crop, and makes a compact tree.

Shropshire Prune. Though for flavour its fruit is the most outstanding of all, it makes but a small, slender tree, and crops only lightly unless planted in a heavy loam, and well supplied with nitrogen. It bears a large, oval fruit at the end of September, which is suitable to use as dessert when fully ripe. Also known as Westmorland damson.

BULLACES

Langley Bullace. Making a compact, upright tree it ripens its fruit later than the damson, as do all the bullaces, a wild species of the prunus family, and which make tough, thorny wood, ideal for hedgerow planting. This is a more recent introduction and possesses the true damson flavour, unlike the others. It is extremely hardy and ripens its fruit about November 1st, but will hang until almost the month end.

New Black Bullace. This is an improvement on the old Black variety, well known to Tudor gardeners. It makes a neat, upright tree, is hardy and bears a tremendous crop of juicy, but acid fruit, best used for jam.

DAMSONS AND BULLACES

Shepherd's Bullace. Valuable for a cold garden, in that it blooms late and bears a heavy crop at the beginning of October, the fruit being grass green when ripe, tart, but juicy.

MYROBALAN OR CHERRY PLUM

Like the damson and bullaces, extremely hardy and bearing a heavy crop of small fruits, suitable for jam and bottling. Obtainable in the red and yellow form, in bloom in March and being almost immune to frost, the fruits are ready for gathering at August Bank Holiday, and are quite delicious when used for flans and tarts, but are quite tasteless and acidy when uncooked. Well worth growing for their blossom too.

Damsons for succession—

River's Early Damson	Early August
Farleigh Prolific	Early September
Bradley's King	Mid-September
Shropshire Prune	End-September
Shepherd's Bullace	Early October
Merryweather Damson	Mid-October
New Black Bullace	Mid-October
Langley's Bullace	Early November

CHAPTER XXIV

THE CHERRY

Their use in the small garden—Soil requirements—Rootstocks—Pollination—Varieties—Acid or culinary Cherries.

THE CHERRY, unless planted for its blossom, as well it might be, is rarely grown in the amateur's garden. For one thing, it succeeds only as a standard, or half-standard, and in this form will take almost ten years to come into reasonable bearing. None of the sweet cherries are able to set any fruit with their own pollen, but it is not enough to plant together several varieties which bloom at the same time, in the expectation that they will pollinate each other, for only certain varieties are capable of pollinating each other. Again, a cherry in the standard form makes such a large tree, that it tends to crowd out other fruit trees growing near, and then again, the question of birds is a constant worry, for even if the trees do set a good crop, quite half the fruit might be taken by birds. By all means plant a cherry, or a number, where space permits, for they remain in bloom longer than any other fruit and provide a charming display during the spring. By planting a wide selection of fruits and beginning with the first of the plums, and ending with the latest flowering apples such as Crawley Beauty and Edward VII, a display of blossom may be enjoyed from the end of March until early June.

But if cherries in the standard form prove too unproductive for the ground they occupy, then the small grower might have room for two or three trees in the fan-shape form, planting them against a wall. It is not suggested that they should be grown instead of pears or plums in this way, though where several outbuildings are available for wall planting, then sweet cherries, the earliest fruit to mature, may be enjoyed in addition to the other fruits. The formation and care of trees has been described in Chapter xx.

SOIL REQUIREMENTS

Cherries like exactly the opposite conditions to the plum, though both are stone fruits. Whereas the plum depends upon a heavy moist soil and copious quantities of nitrogen to crop well, requiring almost no potash and very little lime, the cherry likes a dry soil, preferably a light loam over chalk, a dry, sunny climate, like that of Worcestershire and Kent, and plenty of potash. Nitrogen it does not require in more than average amounts. In the cold districts of the North, and in the warm, but moist climate of the South West, cherries do not crop well, and a too rich soil will also cause excessive gumming, which will eventually weaken the constitution of the tree.

Lime and potash are the primary needs of the cherry, and where planting in a soil deficient in lime, incorporate plenty of lime rubble at planting time. The planting of both plums and cherries is best done during November. When planting take great care to ensure that the bark of the tree is in no way damaged, otherwise it will permit Bacterial Canker, Silver Leaf Disease to enter the wound, plums and cherries being highly susceptible to both diseases. No manure should be given at planting time, nitrogenous manure would only encourage an excess of lush growth, but 1 oz. per tree of sulphate of potash should be given in early April each year. Wood ash, rich in potash may be incorporated at planting time. If planting standard trees, allow them between 20–25 ft., for they form large, spreading heads.

As to pruning, the same remarks apply to the cherry as to the plum, cut out during late spring any dead wood and leave it at that.

ROOTSTOCKS

For centuries, cherries have always been grown on the wild cherry stock, propagated by layering. A form, specially selected by the East Malling Research Station, to give greater uniformity of performance is now being used by nurserymen. Propagation is by budding as described for the plum.

POLLINATION

As previously mentioned, the correct pollination of cherries is a most complicated business, only certain groups being able to pollinate each other, and the research done in recent years to determine the most suitable pollinators would have revolutionised cherry growing, if other conditions were also in favour of their being more widely grown.

It has been carefully noted that each variety has a flowering period of 18 days, almost twice that of the plum, whilst the time from the first to bloom, Nutberry Black, until the latest has finished flowering, Bradbourne Black, is 24 days, again almost twice the flowering period of the plum. Except for the very earliest and latest to bloom, all the cherries overlap with their flowering times on account of their long period of bloom, and yet contrary to expectations, this plays little or no part in their pollination. The sweet cherries will not pollinate the acid or Morellos, and only certain varieties will pollinate each other—

Variety	Pollinators
Amber Heart	Bigarreau Napoleon, Governor Wood, Roundel Heart.
Bigarreau De Schreken	Bradbourne Black, Florence, Gaucher, Roundel Heart.
Bigarreau Napoleon	Bradbourne Black, Florence, Roundel Heart.
Bradbourne Black	Bigarreau Napoleon, Gaucher, Roundel Heart.
Early Rivers	Bigarreau de Schreken, Emperor Francis 'A', Governor Wood, Noir de Guben, Waterloo.
Emperor Francis	Bigarreau de Schreken, Early Rivers, Frogmore, Waterloo.
Florence	Bigarreau Napoleon.
Governor Wood	Early Rivers, Emperor Francis 'A'.
Knight's Early Black	Bigarreau de Schreken, Waterloo.
Roundel Heart	Amber Heart, Bigarreau Napoleon, Bradbourne Black, Governor Wood, Waterloo.
Waterloo	Amber Heart, Bigarreau Napoleon, Florence, Roundel Heart.

Flowering Times—

V.E. = Very early.   V.L. = Very late.

Early.
Early Rivers (V.E.)   Merton Premier
Emperor Francis (V.E.)   Notberry Black (V.E.)
Merton Bigarreau   Waterloo.

Mid-season.
Elton Heart   Knight's Early Black
Frogmore   Merton Heart
Governor Wood   Roundel Heart.

Late.
Amber Heart   Florence Heart (V.L.)
Bigarreau Napoleon   Gaucher
Bradbourne Black (V.L.)   Noble (V.L.)

These are the most reliable pollinators—
- Bigarreau Napoleon
- Roundel Heart
- Waterloo
- Waterloo
- Early Rivers
- Emperor Francis

These will not pollinate each other—
- Early Rivers with Knight's Early Black
- Elton Heart with Governor Wood
- Frogmore with Waterloo
- Noble with Florence Heart

VARIETIES

These are a dozen of the most popular sweet cherries—

Amber Heart. Also pollinated by Waterloo and ready mid-July. This is the best all-round cherry in cultivation, being hardy, a consistent cropper, and does well as a standard or fan-trained tree. Widely used by the canners. The attractive yellow fruits are flushed red. This is also the popular White Heart sold by the barrow boys. If I grew cherries commercially, Waterloo and Amber Heart would be my choice.

Bigarreau Napoleon. Pollinated by Waterloo and Roundel Heart, this is a delicious cherry for dessert, large, very sweet and vivid red in colour. Ready end of July.

Bradbourne Black. Plant in a large garden or orchard, for it makes a large, spreading tree. It is a heavy cropper, the huge crimson-black fruit being of delicious flavour. Excellent for a frosty garden planted with Napoleon and Roundel Heart, for they pollinate each other and all bloom late.

Early Rivers. This is the earliest variety to fruit, ready mid-June and bears huge jet black fruit in profusion. It is a hardy variety, and the tree has enormous vigour.

Emperor Francis. Grown with Early Rivers (or Waterloo) this variety would ensure a crop in June and another (Emperor Francis) in late August. It is a fine all-round variety, the large, dark crimson fruits being of excellent flavour. The first cherry to flower and the last dessert cherry to fruit.

Florence. Another bright red cherry which does well when planted with Napoleon, cropping about ten days later. A heavy cropper as either a standard or fan-trained tree.

Frogmore. Useful for a small garden in that it makes a compact, upright tree. It bears heavily and comes into bearing earlier than most cherries, bearing large yellow and red fruit.

Governor Wood. It makes a large, spreading tree, and pollinated with Emperor Francis or Early Rivers, bears a huge crop of yellow cherries, flushed pink, and being exceptionally rich and sweet.

Knight's Early Black. Of compact habit and useful for a small garden, it bears a heavy crop of large, jet black fruit of excellent flavour.

Merton Heart. This new cherry is now being widely planted to follow Early Rivers and Waterloo. It is a heavy and consistent cropper and bears a large deep-crimson fruit of rich flavour. Should be grown with Emperor Francis or Early Rivers.

Roundel Heart. This variety may also be planted with Waterloo, as they pollinate each other. It produces very large deep-purple fruit which is ready for picking early July.

Waterloo. Early to mid-season flowering, and a suitable pollinator for Early Rivers, Emperor Francis, etc. It makes a compact tree, but bears less regularly than most cherries, though its fruit, deep crimson coloured, is sweet and juicy.

Cherries in order of ripening their fruit—
    Early Rivers—Mid-June
    Governor's Wood—Late June
    Knight's Early Black—Early July
    Frogmore—Early July
    Roundel Heart—Early July
    Merton Heart—Mid-July
    Waterloo—Late July
    Amber Heart—Late July
    Bigarreau Napoleon—Early August
    Florence—Mid-August
    Bradbourne Black—Mid-August
    Emperor Francis—Late August.

To protect the fruit from birds, close mesh fish netting should be hung over the trees as soon as the fruit has set, and this may also be used for covering the heads of young standard trees until they become too large. Some protection for the fruit may then be given by fastening tobacco tin lids together and suspending them amongst the trees, to clatter in the wind.

ACID OR CULINARY CHERRIES

The acid cherries, Flemish Red, Kentish Red and the Morello cherry are all self-fertile and will set their fruit without the aid of a pollinator. They are more hardy than the sweet cherries, their blossoms being less susceptible to frost. For a cold garden they are a valuable fruit, beautiful when in bloom and their leaves take on the autumnal tints of the pear. Both the Flemish and Kentish Red bear their fruit very early, at the end of June, when they are valuable for jam and for tarts; the Morello's fruit is ripe in August. Trees of the Flemish Red are more upright and less drooping than those of the Kentish Red; whilst the Morello makes a densely branched tree. Each may be planted against a cold North wall as a fan tree, or they may be used, with damsons, for a windbreak, or for those cold corners to which other fruit will not take kindly.

The fruit of the Morello is large and takes on a rich crimson-black appearance when fully ripe, but always maintains its acid-bitterness.

CHAPTER XXV

SPRAYING FOR PESTS AND DISEASES

The Apple; Diseases and pests—The Pear; Diseases and pests—The Plum and Damson; Diseases and pests—The Cherry; Diseases and pests.

THE COMMERCIAL grower of fruit becomes more and more bewildered by the constant stream of new insecticides introduced each year by chemical firms, some deadly poisonous and which have caused death to their users, so that the control of pests and disease has become to be recognised as of greater importance than the actual growing of the trees. The modern chemist enjoys a national prestige whilst the fruit grower remains a nonentity. This is not of great importance in itself, but what does give rise to concern is the author's belief that many of the modern stomach upsets, ever on the increase is in part due to the use of so many toxic chemicals, used to control pests and disease on agricultural crops of every description. Perhaps one day we shall have the preparation to control all troubles, and the consumer will also be exterminated. The amateur is fortunate in that he knows what has been used on his fruit trees, but even he will be advised to follow a simple programme, whereby the health of the trees will be maintained without sacrificing valuable pollinating insects and harming his stomach.

THE APPLE

DISEASES

Brown Rot. Control here is difficult, though spraying as for scab seems to give a measure of control. This is a most troublesome fungus in certain areas, the fruit being completely rotten in the centre, while still appearing sound from the outside. Blossom Wilt is another form of the disease and may be known to be present when the blossom on the spurs turns brown and dies back, and it sometimes happens that the whole branch is affected. The variety Lord Derby is often worried by brown rot disease. Late winter spraying with a tar-oil wash gives satisfactory if not complete control.

Canker. This fungus disease generally attacks where scab has already made its presence felt. It is the result of badly drained soil or too heavy applications of nitrogen making for a soft tree. It is observed as reddish coloured bodies clustered together on parts of the wood, which often results in a branch dying back above the canker attack. Most of those varieties resistant to scab frequently succumb to canker and Worcester Pearmain may be included in the list. In those areas where the trouble is frequently seen, one of those highly resistant varieties, Gladstone or Grenadier, should be planted.

The most satisfactory method of eradicating the trouble is to cut away the cankered portion of the wood, and then to apply a dressing of ' Medo,' which penetrates the decayed tissues and so destroys the disease.

Mildew. This is prevalent on old, neglected trees, the shoots taking on a white powdery appearance, and buds which are affected often fail to develop. Spraying with lime-sulphur or Murfixtan as for Scab, will give a certain control, but the new American fungicide, Katharane, used as per maker's instructions, is proving most efficient, but is yet new to this country.

Scab. This is the most troublesome of all apple diseases and frequently attacks trees which suffer from potash deficiency, or have received too heavy supplies of nitrogen, causing a soft, sappy tree. The trouble with the disease is that it attacks all parts of the tree—buds, shoots, leaves and even the fruits, causing the formation of blackish blisters; this not only causes the fruits to rot, but opens up the tree for a host of other diseases. Control may be achieved by spraying with a two per cent strength lime-sulphur solution just before the buds begin to open. Lime-sulphur should not be used on Stirling Castle, St. Cecelia, Lane's Prince Albert, Cox's Orange Pippin, Beauty of Bath, Newton Wonder, Rival, Belle de Boskoop and Egremont Russet, but of these varieties all except Cox's and Newton Wonder are highly resistant to scab.

Another method is to dust the blossom, when in full bloom with Sulfadu Sulphur Dust, the time for applying the dust being governed by the flowering periods. The sulphur-shy varieties should instead be sprayed with Murfixtan, at the rate of 2 pints per 100 gallons of water, applying just as the buds begin to burst. This is an excellent preparation for use with orchards of mixed varieties, not only where planted with different varieties of apple, but also with pears.

THE APPLE
PESTS

Blossom Weevil. Attacking the buds as they open during early summer, the grubs so eat into the stamens that not only do large numbers of blossoms fail to open, but many that do so prove incapable of carrying pollen. The most effective control is to spray the trees during early March with a petroleum-oil emulsion to which is added D.D.T. for control of the Blossom Weevil. Never use petroleum oil on Cox's or Newton Wonder.

Codling Moth. This pest is responsible for the maggoty condition of matured fruit, and is really a serious pest. Its presence is indicated by a pile of brown dirt at the entrance hole of the attacked apple, and upon inspection the fruit will be found to be riddled with holes right through to the core. The moth lays its eggs during June and July following on the period of sawfly attacks and frequently uses the skin of the young fruit for its egg-laying. Spraying with a Derris preparation will ensure almost complete control, and is safer to use than the older, but efficient, lead arsenate and white-oil emulsion, poisonous to humans and which should not be used closer than 8 weeks to the fruit maturing. Lead arsenate must never be used on Beauty of Bath, Miller's Seedling or Grenadier.

Greenfly. In this section, Rosy Aphis is the most troublesome pest. This feeds on the young shoots and even on the fruit as it is forming. Early in winter the aphis lays its eggs on the spurs, but an application of a tar-oil wash during mid-winter will quickly kill these eggs, and a repeat of the spraying the following year should rid the trees of aphis for a number of years. The spraying programme should be so regulated that control measures for red spider should take place immediately after the aphis are destroyed, otherwise the red spider will have the tree to itself and multiply accordingly.

Red Spider. The mites feed on the sap, the young being produced in webs on the backs of the leaves, which eventually turn brown and growth of the tree is halted. Spraying with lime-sulphur, for those varieties which are not sulphur-shy, just after the blossom has fallen and again 3 weeks later will keep the pest under control; or applications of Derris dust at monthly intervals will also give control.

This pest is generally most troublesome with trees growing against a wall, and especially during a period of drought. Frequent syringing of the trees during the period June to September, will help to keep the pest under control. It is also more prevalent in the dryer climate of the East than in the West.

Sawfly. Here the fly lays its eggs in the blossom, the grubs later eating their way inside, then passing from fruit to fruit. Control may be achieved by dusting either with nicotine or derris, a fortnight after the petals have fallen, and repeating this a month later. The preparation Gammexane is also highly effective, and especially so under cooler conditions.

Tortrix Moth. Though not so troublesome as the Codling Moth, Tortrix caterpillars can cause a certain amount of damage by eating into the swelling buds early in spring, and even into the fruit as the summer advances. Their presence may be detected by a spinning together of the young leaves in early summer. Control measures are the same as for the Codling Moth.

Winter Moth. Using their powers of destruction in much the same manner as the Tortrix Moth, attacks from the Winter Moth may be kept at a minimum by grease-banding all trees in October-November, for they hibernate in the soil around the trees and make their way up the stems during mid-winter.

Woolly Aphis. The small brown insects covered in a white cocoon-like substance, attack the bark of the trees, feeding on the sap, like red spider, and causing swelling of the bark which sometimes split and so become an entry for disease. It may be controlled by washing the trees with a preparation called Lindex, using 1 oz. to 2 gallons of water whilst the trees are in the green bud stage.

THE PEAR
DISEASE

Scab. Though the symptoms are the same, scab on pears is a totally different disease from that which attacks apples. Unlike apples, most varieties of pears will not tolerate even dilute lime-sulphur sprays, but again unlike the apple, are tolerant of Bordeaux Mixture, which will control the disease if applied early in June and again when the blossoms have set. Fertility and Doyenne du Comice are frequently troubled by Pear Scab.

For other disease and their treatment see under 'A P P L E.'

THE PEAR
PESTS

Midge. Not often troubled by the sucking insects which attack the apple, this is by far the most troublesome pest. Like the apple sawfly, the pest lays its eggs in the blossoms, the grubs following the same process as those of the sawfly. Both Laxton's Superb and Conference are amongst the most resistant varieties. The pest may be easily controlled by dusting with D.D.T. when the blossom is open, and again a month later. As D.D.T. will also control Capsid, Sawfly and Blossom Weevil, it may be considered the pear grower's great standby, as well as being effective on apples. The treatment for all pests which attack pears is the same as for those which attack the apple.

THE PLUM AND DAMSON
DISEASES

Brown Rot. There are three serious troubles in this section.
(a) Spur Blight.
(b) Blossom Wilt.
(c) Fruit Rot.

(a) Spur Blight is caused by infection of the new leaves which are attacked by spores which later travel down the spurs and on to the branches, causing decay and all that goes with it.

(b) Blossom Wilt is the same infection, only it is the blossom that is mainly attacked, causing it to fall away before the fruit is set.

Both these diseases may be controlled by spraying with a one per cent lime-sulphur solution just before the blossom opens. A wash with petroleum-oil solution in February will also give additional control and the two should keep down Red Spider and destroy the eggs of the leaf-curling aphis. This spraying programme given once every three years, or whenever necessary, should be sufficient to keep the plum orchard in a healthy condition, with a derris spray immediately after petal fall.

(c) Fruit Rot. This is caused by a fungus disease attacking the individual fruits, causing them to mummify on the trees. There is no known cure, at least I do not know of one, but luckily the trouble is not common.

Canker. The varieties Victoria and Czar seem most prone to attack which concentrates on the main stem of plums, causing the troubled area to become decayed. Should the trouble extend completely round the stem, the tree will die back and be of no further use. Myrobalan B. stock has proved very resistant.

Silver Leaf. This fungus is the most dreaded of all plum diseases, entering the tree through a cut or break which is the primary reason why all pruning should be done during May. From June to September the plum exudes a gummy substance wherever a cut has occurred and this will tend to keep out the fungus. Where breakages occur at other times of the year, the wound should be treated with white lead paint as a precaution against the fungus making entry. The fungus lives on the dead wood which must be removed throughout the life of the tree. A Ministry of Agriculture Order actually makes this compulsory by mid-July of each year, when all dead wood must be removed and burnt.

THE PLUM AND DAMSON
PESTS

Leaf Curling Aphis. This greenfly causes damage to the leaves and young fruit in much the same manner as the Red Spider. The eggs are laid on the branches in late autumn and hatch out in early spring. A tar-oil wash given during December or January, will kill off all eggs and control is therefore not difficult.

Plum Maggot. This may prove troublesome in some seasons, the eggs being laid on the young fruit from whence the grubs work their way into the centre. Fruits will either fall prematurely or will be uneatable if they mature. The best method of control is to soak the trunks of the trees with Mortegg during winter, and to dust the open blossom and young fruit with D.D.T.

Sawfly. The Plum Sawfly lays its eggs in the flowers where the hatched grubs remain until the fruit begins to form, when they begin their tunnelling in much the same was as in apples and pears. Effective control may be made by spraying with Lindex, 1 oz. to 2 gallons of water, immediately after petal fall.

THE CHERRY
DISEASES

Canker. Just as it affects the plum so does the cherry suffer from the disease, entering wounds of the tree over the autumn and winter months when gumming does not act as a deterrent. Pruning should take place early in summer and, prevention being better than cure, the whole tree should be sprayed during early winter with a solution of Bordeaux Mixture.

Leaf Scorch. When the leaves change to a mottled green and yellow colour and remain on the trees long after the period when they should have fallen, leaf scorch disease will be the cause. Luckily cherries will tolerate Bordeaux Mixture, and an application should be given just before the buds open. The same treatment will also rid the tree of the spores of Brown Rot Blossom Wilt disease which can cause serious damage to the culinary or acid cherries.

THE CHERRY
PESTS

Black Fly. The tiny black eggs winter on the twigs and if not killed by a January tar-oil spray will hatch out minute grubs early in summer, which will devour not only the leaves, but much new growth.

Winter Moth. As described for apples, this pest lives in the soil beneath the trees and will crawl up the trunk during the early winter months to lay their eggs on the twigs and branches. Grease banding of the trees at the end of Otcober as for apples, and dusting the foliage with D.D.T. in early summer should prevent any serious attack.

Simple precautions such as grease banding, tar-oil washing in the early winter months and the occasional use of derris powder and D.D.T., should be all that is necessary for the amateur. Much more important is to plant healthy trees and to maintain their vigour by regular attention to manurial requirements.

INDEX

A

Apples, 13
  blossom of, 40, 118
  climate, 13, 26
  crab, 119
  diseases of
    brown rot, 61, 148
    canker, 149
    mildew, 39, 149
    scab, 26, 39, 149
  for a chalk soil, 20
  for a clay soil, 22
  for a wet soil, 23
  grafting, 65
  harvesting, 75
  in grass, 28
  in pots, 114
  in tubs, 11, 67, 114
  its introduction, 9
  pests
    Blossom Weevil, 67, 150
    Codling Moth, 150
    Red Spider, 150
    Sawfly, 151
    Tortrix Moth, 151
    Winter Moth, 151
    Woolly Aphis, 32, 34, 151
  planting, 44
  pollination, 9, 16, 37
    diploid varieties, 37
    triploid varieties, 37, 39
  protection of, 17
  pruning, 58
  renovating, 60
  rootstock of, 31, 47
    most robust, 34
    semi-dwarf, 32
    very dwarf, 31
    vigorous, 33
  spur-bearing, 21, 51, 71
  staking and tying, 48
  storage of, 11, 77
  thinning, 15
  tip-bearing, 21, 51, 70, 100
  Varieties
    Acme, 88
    Adam's Pearmain, 22, 24, 26, 32, 94, 116
    Allington Pippin, 16, 22, 24, 38, 91
    Annie Elizabeth, 15, 17, 76, 107, 119
    Arthur Turner, 73, 105, 119
    Autumn Pearmain, 91
    Barnack Beauty, 21, 24, 76, 99, 100
    Barnack Orange, 21, 24
    Beauty of Bath, 11, 29, 39, 40, 78, 96, 149
    Belle de Boskoop, 82, 149
    Blenheim Orange, 10, 16, 26, 32, 63, 72, 98
    Bowden's Seedling, 76
    Bramley's Seedling, 9, 13, 14, 23
    Brownlee's Russet, 75, 103, 118
    Celia, 86
    Charles Ross, 21, 24, 26, 81, 84
    Christmas Pearmain, 40, 76, 92
    Claygate Pearmain, 76, 92, 116
    Cockle's Pippin, 102
    Cornish Aromatic, 16, 93
    Cornish Gilliflower, 11, 94, 100
    Cottenham Seedling, 119
    Court Pendu Plat, 16, 17, 94
    Cox's Orange Pippin, 9, 10, 13, 29, 32, 37, 38, 40, 93, 149
    Crawley Beauty, 16, 17, 38, 40, 78, 83, 84, 119
    D'Arcy Spice, 13, 16, 38, 102
    Devonshire Quarrenden, 11, 16, 91
    Duchess of Oldenburg, 81, 84, 116
    Duke of Devonshire, 76
    Easter Orange, 94
    Edward VII., 16, 17, 22, 38, 40, 77, 83, 84
    Egremont Russet, 37, 38, 39, 101, 116, 149
    Ellison's Orange, 13, 16, 24, 32, 38, 91, 116
    Elton Beauty, 86
    Emneth Early, 73, 104
    Forge, 17, 24, 82
    Franklin's Golden Pippin, 102
    Gascoyne's Scarlet, 21, 24, 82

George Cave, 77, 78, 86
Gladstone, 40, 78, 97, 149
Golden Noble, 16, 106
Golden Pippin, 10
Golden Russet, 103
Gravenstein, 17, 39, 75, 93
Grenadier, 16, 23, 24, 29, 34, 39, 40, 76, 105, 149
Hereford Cross, 87
Herring's Pippin, 22, 24, 92
Heusgen's Golden Reinette, 94
Howgate Wonder, 106
Irish Peach, 90
James Grieve, 13, 14, 15, 16, 37, 38, 39, 40, 97
Kidd's Orange Red, 87
King of the Pippins, 22, 24, 92, 116
King of Tomkin's Country, 94
King's Acre Pippin, 76
Lady Henniker, 82
Lady Sudeley, 11, 15, 90, 116
Lane's Prince Albert, 15, 16, 38, 40, 51, 76, 106, 149
Laxton's Advance, 40, 90
Laxton's Epicure, 40, 91
Laxton's Exquisite, 40, 91
Laxton's Favourite, 87
Laxton's Fortune, 17, 40, 91
Laxton's Rearguard, 77, 89
Laxton's Royalty, 16, 17
Laxton's Superb, 15, 16, 23, 24, 40, 76, 98
Lemon Pippin, 84
Lord Derby, 16, 23, 24, 29, 34, 39, 105, 148
Lord Lambourne, 16, 39, 98
Lord Suffield, 16
Margil, 93
May Queen, 95, 116
Melba, 17
Merton Prolific, 88
Merton Russet, 118
Merton Worcester, 37, 87
Michaelmas Red, 86, 87
Miller's Seedling, 16, 29, 32, 38, 72, 97
Monarch, 16, 23, 24, 106
Mother, 17, 92
Newton Wonder, 15, 16, 17, 22, 29, 38, 39, 63, 72, 76, 107, 149
Ontario, 119
Opalescent, 82, 84
Orlean's Reinette, 102
Pearl, 88, 100
Peasgood's Nonsuch, 106
Pineapple Russet, 101
Pott's Seedling, 22, 24, 105
Powell's Russet, 103
Rev. Wilks, 40, 105

Ribston Pippin, 39, 40, 75, 77, 93
Rival, 98
Rosemary Russet, 102
Royal Jubilee, 16, 17, 38, 40
Sam Young, 23, 24, 102
Shaw's Pippin, 88
Sowman's Seedling, 105
St. Cecilia, 39, 149
St. Edmund's Russet, 38, 100, 101
St. Everard, 21, 38, 40
Stirling Castle, 37, 149
Sturmer Pippin, 17, 38, 75, 95
Sunset, 16, 17, 26, 38, 99, 116
Taunton Cross, 16, 88
Tydeman's Early Worcester, 87
Tydeman's Late Orange, 89
Upton Pyne, 84, 119
Wagener, 40, 83, 84
Wealthy, 82
White Transparent, 81
Winston, 15, 33, 99
Woolbrook Pippin, 16, 83, 84, 119
Worcester Pearmain, 10, 14, 15, 16, 17, 24, 26, 29, 37, 38, 63, 97, 100, 149
Wyken Pippin, 94

B

Biennial cropping, 38, 72
Bud, nicking and notching, 73
  propagation of, 130
  stimulation of, 52, 54
Bullaces, 139
  Varieties
    Langley Bullace, 140
    New Black Bullace, 140
    Shepherd's Bullace, 141
Bunyard, Edward, 92, 102, 137
Bush trees, 51

C

Cherries, Morello, 125, 146
  Diseases
    Brown rot, 125
    pruning, 125
Cherries, Sweet
  blossom, 120, 142
  Diseases
    canker, 153
    leaf scorch, 153
  harvesting, 80
  Pests
    Black Fly, 154
    Winter Moth, 154
  pollination, 143
  protecting fruit, 146
  pruning, 124

INDEX

rootstocks, 143
soil requirements, 142
Varieties
  Amber Heart, 144, 145
  Bigarreau Napoleon, 144, 145
  Bradbourne Black, 121, 144, 145
  Early Rivers, 121, 144, 145
  Emperor Francis, 144, 145
  Florence, 144, 145
  Frogmore, 80, 146
  Governor Wood, 144, 146
  Knight's Early Black, 144, 146
  Merton Heart, 121, 146
  Roundel Heart, 144, 146
  Waterloo, 144, 146
Cordons, 50, 53
  formation, 53
  planting, 46
  pruning, 72
  staking and tying, 48

D

Damson, 139
  for windbreak, 17, 121, 139
  Varieties
    Bradleys King, 139
    Farleigh Prolific, 140
    Merryweather, 140
    River's Early, 140
    Shropshire Prune, 140
De-horning, 61, 71

E

Espaliers, 48, 50, 54
  pruning of, 54
Evelyn, John, 11

F

Frost, 14, 19, 45
  apples for, 15
Fungicides
  Bordeaux mixture, 151, 153
  lime sulphur, 39, 149
  Medo, 58, 149
  Murfixtan, 149
  sulphur dust, 149

G

Grafting, 65

H

Harris, Richard, 10
Hogg, Robert, 11, 14, 23, 24, 102, 137

I

Insecticides
  D.D.T., 150, 152, 154
  derris, 150
  Gammexane, 151
  grease-banding, 151, 154
  lead arsenate, 150
  Lindex, 151, 153
  Mortegg, 153
  petroleum-oil, 150
  tar-oil, 150, 153

J

John Innes Institute, 37

L

Le Gendres, 35
Lime, 23, 25
Lindley, 11, 22, 77

M

Maidens, 47, 50, 54
  their culture, 50
Manures, inorganic
  ammonium sulphate, 129
  ferron sulphate, 30
  magnesium carbonate, 26, 29
  magnesium sulphate, 30
  nitrogenous, 29
  phosphates (bone meal), 27, 29
  potash (sulphate of) 26, 29, 117
Manures, organic (nitrogenous)
  composted straw, 28
  farmyard, 23, 30, 117, 129
  fish manure, 30
  fish waste, 27
  pig, 26
  poultry, 26
  seaweed, 27
  shoddy, 27, 129
Mulching, 27, 28, 30, 49, 129

P

Parkinson, 11, 120
Pear, 18, 35, 108
  blossom, 120
  climate, 18
  Disease
    scab, 19, 151
  foliage, 120
  harvesting, 78
  incompatability, 35
  its origin, 18
  manure requirements, 30
  Pests
    midge, 151
  pollination, 42
  rootstocks, 35
  storing, 79
  Varieties
    Bergamotte d'Esperen, 112, 116
    Beurré Bedford, 109, 116
    Beurré Bosc, 120
    Beurré Clairegeau, 120

INDEX

Beurré d'Amanlis, 35, 43, 65, 109
Beurré Easter, 112
Beurré Hardy, 18, 35, 43, 110, 120
Beurré Superfin, 110
Bristol Cross, 35, 110
Catillac, 18, 19, 112
Clapp's Favourite, 19, 64
Conference, 43, 78, 110, 116, 152
Doyenne d'Eté, 73
Doyenne du Comice, 10, 18, 35, 64, 110
Dr. Jules Guyot, 18, 35, 109
Durondeau, 18, 43, 64, 108, 110
Emile d'Heyst, 110
Glou Morceau, 111
Gorham, 109, 116
Jargonelle, 18, 108, 120
Josephine de Malines, 112, 120
Laxton's Early Market, 108
Laxton's Foremost, 110
Laxton's Record, 111
Laxton's Satisfaction, 111
Laxton's Superb, 18, 79, 108, 109, 116, 152
Louise Bonne, 43, 65, 111, 116
Olivier de Serres, 112
Packham's Triumph, 35, 111
Pitmaston Duchess, 19, 111, 120
Roosevelt, 18, 108, 111, 116
Santa Claus, 111, 116, 120
Seckle, 43
Triomphe de Vienne, 109
William's Bon Cretien, 18, 35, 79, 108, 109
Winter Nelis, 19, 111, 116
wall trees, 30
Peat, 23, 28, 47, 115
Plum, budding, 130
  Diseases
    brown rot, 152
    canker, 152
    silver leaf, 122, 127, 132, 152
  harvesting, 79
  Myrobalan, 141
  Pests
    Leaf Curling Aphis, 135
    Maggot, 153
    Sawfly, 153
  pollination, 127
  pruning, 122
  rootpruning, 123
  rootstocks, 129
  soil preparation, 128
  suckers, 123
  their culture, 126
  Varieties
    Angelina Burdett, 80, 133, 135
    Belle de Louvain, 128, 133
    Black Prince, 132, 136
    Blue Tit, 132
    Bryanston Gage, 128, 137
    Coe's Golden Drop, 80, 127, 129, 134
    Count Althann's Gage, 128, 130, 137
    Czar, 128, 129, 135
    Denniston's Superb, 136
    Early Gage, 137
    Early Laxton, 132, 136
    Giant Prune, 133
    Goldfinch, 132, 136
    Golden Transparent, 138
    Green Gage, 129, 138
    Jefferson, 128, 133, 136
    Kirke's Blue, 129, 134, 136
    Late Transparent Gage, 128, 137
    Laxton's Bountiful, 133, 135
    Laxton's Cropper, 80, 134
    Laxton's Delicious, 80, 134, 135
    Laxton's Gage, 137
    Laxton's Olympia, 135
    Marjorie's Seedling, 127, 130 135
    Monarch, 128, 130, 135
    Oullin's Golden Gage, 128, 133, 135
    Pershore (Yellow), 128, 133, 135
    Pond's Seedling, 128, 129, 134, 135
    President, 128, 130, 135
    Reine Claude de Bavay, 138
    River's Early Prolific, 132, 135
    Severn Cross, 135
    Thomas Cross, 134
    Utility, 133
    Victoria, 129, 133, 135
    Warwickshire Drooper, 128, 134, 135
    White Magnum Bonum, 134
Pre-harvest drop, 78
Pruners, choice of, 56
Pruning, bark, 64
  functions of, 58
  regulated system, 70
  renewal system, 71
  root, 59, 63
  stone fruits, 123
  summer, 72
  vigorous trees, 63
  young trees, 62
Pyramids (dwarf), 50, 52

S

Scion, 46

INDEX

Scott, John, 108, 128
Soil, 20, 45
  acid, 25
  chalk, 19, 20, 129
    manuring, 21, 129
    varieties for, 21
  clay, 22
    varieties for, 22
  nitrogen in the, 28
  sandy, 23
  wet, 23
    varieties for, 23
South Eastern Agricultural College, 14
Standard trees, 51

T
Taylor, H. V., 16, 23
Thompson, Robert, 78, 93